PALAEONTOGRAPHICAL SOCIETY MONOGRAPHS

BRITISH ORDOVICIAN AND SILURIAN PROETIDAE (TRILOBITA)

ROBERT M. OWENS

Pages 1–98 : Plates 1–15

Publication No. 535, issued as part of

Volume 127 for 1973

Recommended reference to this publication:

Owens, R. M. 1973. British Ordovician and Silurian Proetidae (Trilobita). *Palaeontogr. Soc.* [*Monogr.*], 98 pp., 15 pls.

ABSTRACT

British Ordovician and Silurian proetid trilobites are described and referred to 12 genera, 4 subgenera, 36 named species and subspecies (11 new) and 21 other unnamed forms, distributed among the three subfamilies Proetinae, Tropidocoryphinae and Warburgellinae (subfam. nov.).

The genera classified in the Proetidae probably evolved from two major stocks whose origins appear to lie in the early Ordovician hystricurine trilobites.

Les Proetidae (Trilobites) de l'Ordovicien et du Silurien des Îles Britanniques

RÉSUMÉ

Les trilobites Proetidae de l'Ordovicien et du Silurien de Grande-Bretagne sont décrits et rapportés à 12 genres, 4 sous-genres, 36 espèces ou sous-espèces nommées (dont 11 nouvelles) et 21 autres formes laissées en nomenclature ouverte, le tout réparti en trois sous familles: Proetinae, Tropidocoryphinae et Warburgellinae (nouv. sous-fam.).

Les genres rangés dans les Proetidae évoluent probablement à partir de deux stocks principaux dont l'origine semble se situer chez les trilobites Hystricurinae de l'Ordovicien inférieur.

Britische Ordoviz- und Silur-Proetiden (Trilobita)

ZUSAMMENFASSUNG

Die britischen Ordoviz- und Silur-Proetiden werden beschrieben und 12 Gattungen, 4 Untergattungen, 36 benannten Arten und Unterarten (11 neu) sowie 21 unbenannten Formen zugewiesen; sie verteilen sich über die drei Unterfamilien Proetinae, Tropidocoryphinae und Warburgellinae (subfam. nov.).

Die als Proetiden klassifizierten Gattungen entwickelten sich wahrscheinlich aus zwei grösseren Stöcken, deren Ursprung bei den früh-ordovizischen hystricurinen Trilobiten zu liegen scheint.

Британские ордовикские и силурийские проетиды (Trilobita)

РЕЗЮМЕ

Описаны британские ордовикские и силурийские представители сем. Proetidae, отнесенные к 12 родам, 4 подродам, 36 видам и подвидам (11 новых) и 21 форме в открытой номенклатуре. Изученные таксоны входят в состав подсемейств: Proetinae, Tropidocoryphinae и Warburgellinae (subfam. nov.). Рассматриваемые проетиды, возможно происходят от двух основных стволов, которые берут начало от ранне-ордовикских хистрикурин.

Made and printed in Great Britain
Adlard & Son Ltd., Bartholomew Press, Dorking

BRITISH ORDOVICIAN AND SILURIAN PROETIDAE (TRILOBITA)

CONTENTS

INTRODUCTION AND ACKNOWLEDGEMENTS

Trilobites classified in the family Proetidae Salter, 1864, occur in rocks ranging in age from Ordovician (Arenig) to late Permian; they are world-wide in their distribution and achieve their maximum abundance and diversity in the Devonian. Devonian species have been the subject of the greater proportion of research carried out on proetids, those from central Europe attracting the most attention, having been described and figured by Barrande (1852, 1872), Novák (1890), R. & E. Richter (1912–56), Přibyl (from 1946), Erben (from 1951) and Alberti (from 1962). Elsewhere Devonian proetids have been described in recent years from North America by Stumm (1953) and by Ormiston (from 1967), from Siberia by Yolkin (from 1965), from Morocco by Alberti (from 1964) and from south-west England by Selwood (1965). Carboniferous proetids have likewise received much attention, especially in more recent years by Osmólska (from 1962) and by G. & R. Hahn (from 1964). Conversely, pre-Devonian proetids have been largely neglected, and what little information there is on them can only be gleaned from perusal of a large number of scattered references. The main reasons for their having been rather overlooked are that specimens are usually uncommon, small and fragmentary, and are inconspicuous fossils. Proetids generally form only a minor element of most Lower Palaeozoic trilobite faunas, and little attempt has been made to classify them on a rational basis. Species have been commonly assigned to a limited number of 'convenient' genera. The most important references to date on British Lower Palaeozoic proetids are to be found in papers by Nicholson & Etheridge (1879), Reed (1904, 1914, 1935, 1940), Begg (1939, 1947, 1950, 1951), Tripp (1954, 1967), Bancroft (1949), Whittard (1961, 1966), Dean (1962, 1963), Whittington (1966a), Temple (1969, 1970) and Ingham (1970). Only a limited number of species is well illustrated, in the more recent papers, and most species are badly in need of revision.

In order to achieve a fuller understanding of the phylogeny of the Proetidae it is essential to have more information on the early species, and to this end I have been undertaking a revision of the north European Ordovician and Silurian forms. This monograph, together with two papers on Scandinavian proetids (Owens 1970, 1973) completes a survey of some three-quarters of these forms.

The work for this monograph was commenced in 1967 at Leicester University under the supervision of Dr. J. H. McD. Whitaker and Professor P. C. Sylvester-Bradley, and I greatly appreciate their guidance and encouragement. Through the kindness of Professor Dr. H. K. Erben, and with the aid of a scholarship from the Rheinisch-Friedrichs-Wilhelms Universität I was able to spend a year at the Institut für Paläontologie in Bonn, under the expert supervision of Professor Erben and Dr. W. Haas. I am deeply grateful also for funds arranged by Mr. R. J. K. Owens, a grant from the Charles Henry Foyle Trust, Birmingham, and a scholarship from the University of Leicester.

In addition to those named above, I have had the benefit of useful discussion with Professor H. B. Whittington, F.R.S., Dr. D. L. Bruton, Dr. J. K. Ingham, Dr. V. Jaanusson and Dr. A. W. A. Rushton. Much of the material used in this monograph is from museum collections, and I have pleasure in thanking the following for kindly placing collections at my disposal (abbreviations in parentheses are those employed for the various institutions mentioned throughout the text; an alphabetical list precedes the descriptions of the Plates): Dr. W. T. Dean and Mr. S. F. Morris, British Museum (Natural History), London (**BM**); Dr. A. W. A. Rushton, Institute of Geological Sciences, Geological Survey and Museum, London (**GSM**); Dr. R. B. Rickards and Dr. C. L. Forbes, Sedgwick Museum, Cambridge (**SM**); Mr. J. M. Edmonds and Mr. H. P. Powell, Oxford University Museum (**OUM**); Dr. I. Strachan, Department of Geology, Birmingham University (**BU**); Dr. M. G. Bassett, National Museum of Wales, Cardiff (**NMW**); Dr. J. K. Ingham, Hunterian Museum, Glasgow (**HM**); Mr. G. F. Wilmott, Yorkshire Museum, York (**YM**); Dr. C. O'Riordan, National Museum of Ireland, Dublin (**NMI**); Dr. M. Muir, Murchison Museum, Imperial College of Science and Technology, London (**ICMM**); Dr. A. Lord, University of Hull (**HUR**—Rickards

Collection, **HUD**—Ingham Collection); Dr. W. Struve, Senckenberg Museum, Frankfurt-am-Main, Germany (**SMF**); Dr. V. Zazvorka, National Museum, Prague, Czechoslovakia (**NMP**); Dr. V. Jaanusson, Naturhistoriska Riksmuseet, Stockholm, Sweden (**RM**); Dr. J. Bergström, University of Lund Palaeontological Institute, Lund, Sweden (**LPI**); Dr. S. Stuenes, University of Uppsala, Palaeontological Institute, Uppsala, Sweden (**UM**). In addition, the following individuals have generously allowed me to use specimens from their personal collections (now housed in above institutions unless otherwise stated): Dr R. Addison; Miss M. Breeze; Dr. L. R. M. Cocks; Dr. J. C. Harper, Liverpool University (**LU**); Professor C. H. Holland; Mr. M. D. Jones, Leicester City Museum (**LCM**); Dr. P. D. Lane; Dr. D. Palmer, Trinity College, Dublin (**TCD**); Dr. J. S. W. Penn; Dr. R. B. Rickards; Dr. M. Romano, Sheffield University; Dr. J. H. Shergold; Mr. Derek Siveter; Miss J. Vinnicombe; Dr. J. H. McD. Whitaker, Leicester University (**LRU**); Dr. A. M. Ziegler.

Professor H. B. Whittington, F.R.S. and Dr. M. G. Bassett kindly read a draft of the manuscript and made a number of helpful suggestions for its improvement. Mrs. K. Barrett drafted the tables and the lettering on the other text-figures.

METHODS OF STUDY

The main drawback of museum material is that many specimens in older collections are only vaguely localized. These have been supplemented to some extent by personal collecting, and by accurately localized collections made by various people (see above) involved, for example, in mapping projects.

All specimens have been photographed by the author with a Leitz Aristophot camera and illuminated with a fluorescent ring light. Before photographing, most were given a coating of dilute black 'opaque' to give even contrast and all were given a light coating of ammonium chloride sublimate. Some photographs of minute growth stages were made on the scanning electron microscope at the Geology Department, University of Leicester, by G. L. C. McTurk, to whom I am most grateful.

The measurements used are shown on Text-fig. 1, and were made with a Vernier caliper on larger specimens and with a micrometer eyepiece on smaller ones. Measurements, all in millimetres, are recorded for most of the figured specimens and some additional ones, so that a series is given over the known size range. Estimated measurements are given in parentheses, and 'E' and 'I' refer to the external surface and internal moulds respectively, and 'E/I' to partially exfoliated specimens. No measurements are given for highly distorted specimens as these would serve no value in comparisons.

For illustration and all measurements, specimens were orientated so that on the cephalon the palpebral lobe was horizontal, and on the pygidium the axial furrows were horizontal. Some of the 'code letters' ('K', 'Z', 'Y', 'W' and 'X' for measurements of glabellar width, pygidial length, pygidial axial length, pygidial width and pygidial axial width respectively) used by Hughes (1969, pp. 51–5) have been employed, but 'A' is used for all cephalic sagittal length measurements. Definition of measurements is shown in Text-fig. 1.

Terminology used herein is essentially that of Harrington *et al.* (*in* Moore 1959, pp. O117–26), but includes additional terms used by Owens (1970, p. 309) and the following additions and modifications. The term **tropidium** was first proposed by R. & E. Richter (1919, p. 3) for a ridge which traverses the preglabellar field and free cheek a short distance inside and parallel with the cephalic border. It has been used in this sense by Harrington *et al.* (*in* Moore 1959, p. O126), who called it a "tropidia". Whittard (1938, p. 97) used the term in an entirely different sense, and considered it to represent a sagittally expanded anterior border, as developed in *Cyphoproetus binodosus* (Whittard) (see Pl. 6, fig. 12). Ormiston (1967, p. 33) stated that "the true tropidium corresponds in position with the inner edge of the doublure, and is therefore a totally different structure from the ridge on the genal field and fixed cheek of many proetids that has been called a tropidium by some authors". The tropidium in Richter's sense, however, does not correspond

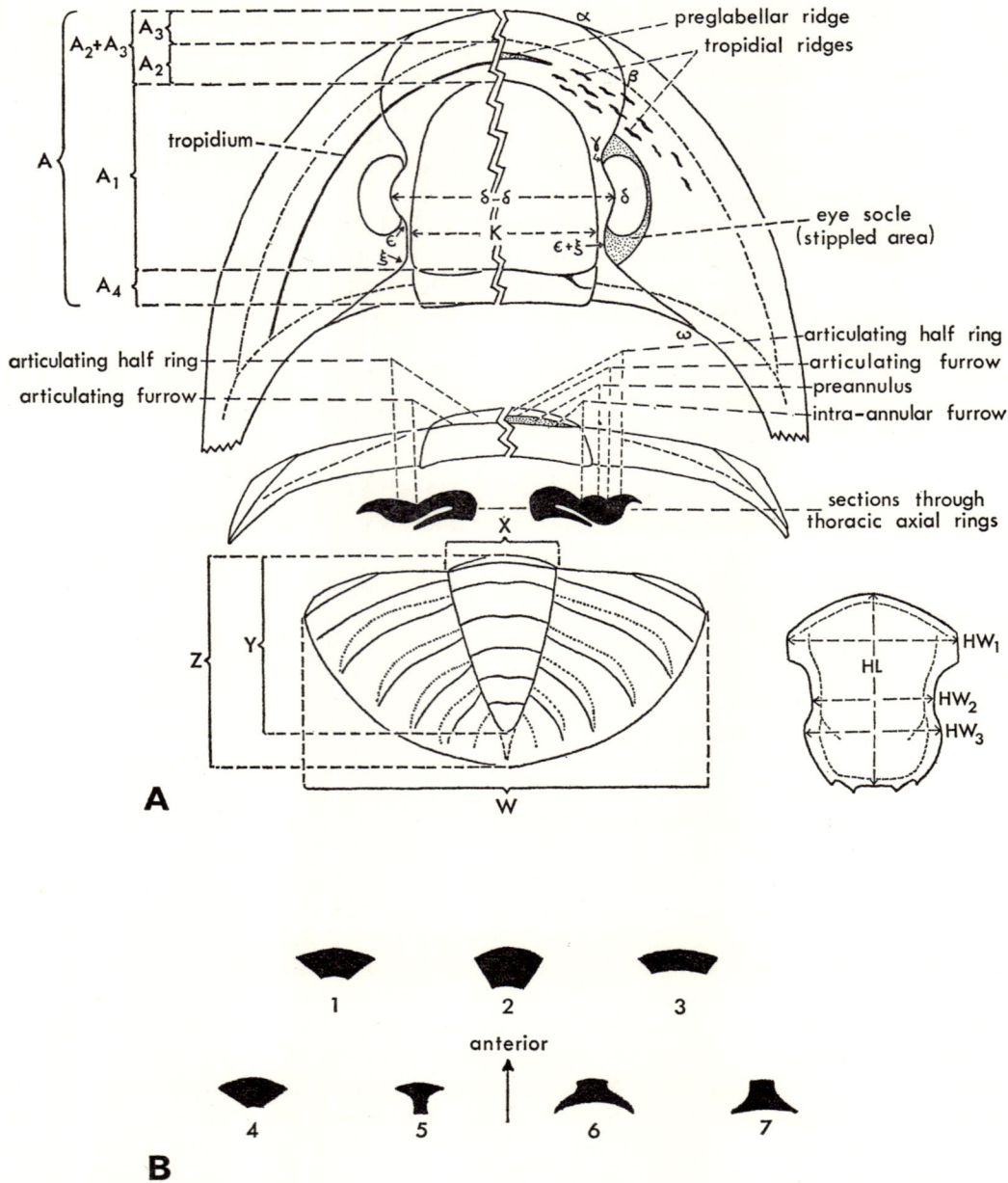

TEXT-FIG. 1. Proetid morphology and measurements.

A. A, A_1, A_2, A_3, A_4=sagittal lengths of cephalon, glabella, preglabellar field, anterior border and occipital ring respectively; A_2+A_3=combined sagittal lengths of preglabellar field and anterior border; K=maximum transverse glabellar width: $\delta-\delta$=distance between abaxial extremities of palpebral lobes; HL=sagittal length of hypostome; HW_1, HW_2, HW_3=transverse widths across regions of anterior wings, 'waist' and posterior wings respectively; Z, Y=sagittal length of pygidium and pygidial axis respectively; W, X=maximum transverse widths of pygidium and pygidial axis respectively.

B. A selection of proetid rostral plates. 1, *Proetus (Proetus) concinnus*; 2, *Cyphoproetus binodosus*; 3, *Astroproetus reedi*; 4, *Decoroproetus jamesoni*; 5, *Stenoblepharum warburgae*; 6, *Warburgella rugulosa canadensis*; 7, *Prantlia grindrodi* (5 after Owens 1973, fig. 9E; 6 after Ormiston 1971a, pl. 19, fig. 7; the others based on specimens figured herein). 1–5 show a range of trapezoidal rostral plates in which the connective sutures converge backwards, the examples being taken from the Proetinae (1, 2) and from the Tropidocoryphinae (3–5). 6 and 7 have the connective sutures diverging backwards, a condition found only in the Warburgellinae.

with the inner edge of the doublure which only extends as far as the cephalic border furrow. The term is thus used here as Richter originally defined it, but used only when one or more *continuous* ridges are referred to, as in *Warburgella* (*Warburgella*) *stokesii* (Murchison), as seen in Pl. 13, fig. 5c (singular: tropidium, plural: tropidia). Short discontinuous raised ridges occupying a similar position to the tropidium (e.g. in *Decoroproetus scrobiculatus* sp. nov. (Pl. 9, figs. 12, 13, 15, 16) and in *Astycoryphe? junius* (Billings)—see Whittington 1960, pl. 54, figs. 11, 17, 18)—are here given the new term **tropidial ridges**.

That area of the free cheek lying between the eye, and the lateral and posterior border furrows is termed the **field** of the free cheek.

Harrington *et al.* (*in* Moore 1959, p. O125) defined 'preglabellar ridge' as a longitudinal median ridge crossing the preglabellar field. Some warburgellines, e.g. *Warburgella* (*Warburgella*) *stokesii* (Pl. 13, figs. 11, 13b) and *W.* (*Tetinia*) *ludlowensis* (Alberti) (Pl. 14, figs. 16a, 17), have a transverse ridge crossing the preglabellar field a short distance behind the anterior border furrow. This is referred to herein as a **transverse preglabellar ridge** to avoid confusion with the longitudinal feature of Harrington *et al.*

The longitudinal section of the preglabellar field is described as **sigmoidal** herein when the profile is convexo-concave, i.e. convex immediately in front of the preglabellar furrow, then concave as far as the anterior border furrow (e.g. Pl. 8, fig. 1b; Pl. 9, fig. 3c).

Three terms—'imbricate', 'flat-topped' and 'scalloped' are introduced to describe the three important types of pygidial pleural rib-structure found in proetids (see Text-fig. 2). These terms are defined as follows. **Imbricate** type: exemplified by *Decoroproetus* and well seen in Tropidocoryphinae. Pleural furrow with steep anterior slope which increases in depth abaxially, and shallow

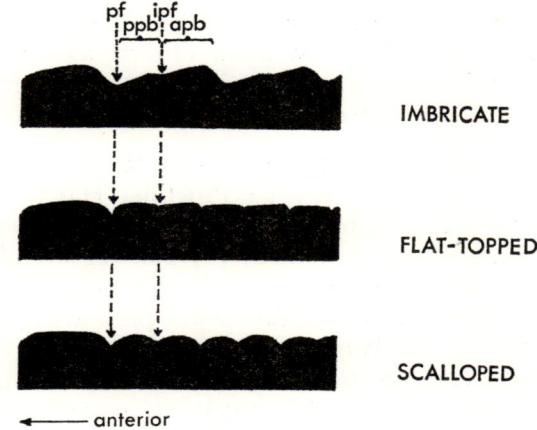

TEXT-FIG. 2. Principal types of proetid pygidial pleural ribs.
The sections are drawn parallel to the lateral margin, approximately half way across the pleural area. There has been some vertical exaggeration to emphasize the structure. Key: pf=pleural furrow; ipf=interpleural furrow; apb=anterior pleural band; ppb=posterior pleural band. *Decoroproetus* is 'imbricate', *Warburgella* is 'flat-topped', *Proetus* is 'scalloped'.

posterior slope. Interpleural furrow commonly weak, with very shallow anterior and posterior slopes. Anterior and posterior pleural bands inclined downwards and forwards. **Flat-topped** type: exemplified by *Warburgella* and confined to Warburgellinae. Narrow pleural furrow with steep anterior and posterior slopes; shallow, narrow interpleural furrow with shallow anterior and posterior slopes. Anterior and posterior pleural bands more or less horizontal. The flat-topped type of pygidial pleural rib-structure is evidently derived from the imbricate type by reduction in inclination and in height of the anterior slope of the pleural furrow and by steepening of the posterior slope. **Scalloped** type: exemplified by *Proetus* and found in the Proetinae and their descendants. Pleural furrows deeper than interpleural, both moderately steep-sided. Anterior pleural band and posterior pleural band gently arched in longitudinal profile. The imbricate type of pygidial pleural rib-structure remains readily identifiable on internal moulds (e.g. Pl. 9, fig. 20), but the flat-topped and scalloped types may not be so readily differentiated in this condition (cf. Pl. 1, fig. 3 and Pl. 14, fig. 11).

I have followed the stratigraphical terminology and correlation employed in the Geological Society of London correlation charts (Ordovician—Williams *et al.* 1972; Silurian—Cocks *et al.* 1971).

Where possible six- or eight-figure National Grid references are given for localities mentioned in the text, but in many cases it is possible to give only rather vague locality information for specimens in old museum collections. For many Llandovery localities in the Welsh Borderland the locality numbers listed by Ziegler, Cocks & McKerrow (1968) are given in square brackets after the National Grid reference—e.g. [ZCM loc. 58].

SYSTEMATIC DESCRIPTIONS

RECOGNITION OF PROETIDAE

On first sight small proetid pygidia can easily be confused with those of small phacopids. However, when internal moulds are available the proetids can be distinguished by the presence of fine, parallel terrace lines on the doublure, a feature not found in phacopids. The trilobites most easily confused with proetids are otarionids, as both possess terrace lines on the doublure and both have an opisthoparian suture. Otarionid pygidia are small, considerably wider than long and have a short axis with few axial rings. Proetid pygidia are proportionally longer, and generally have a larger number of axial rings. The otarionid cephalon is distinguished from proetids by the combination of the convex preglabellar field, the eye well out from the axial furrow, the prominent basal glabellar lobes and the incurved lateral margin at the base of the genal spine. Some proetids possess some of these characters, but apparently none possesses all.

Family PROETIDAE Salter, 1864
(=Prionurides Hawle & Corda, 1847; =Proetiden Hawle & Corda, 1847)

Diagnosis. Glabella tapering forward, ovate or parallel-sided in all but members of Griffithidinae Hupé, 1953, where frontal lobe is expanded; 1–4 pairs of lateral glabellar furrows, which may be incised; eye generally close to glabella, but may be distant from it in genera with reduced eyes; occipital ring with or without lateral lobes; rostral plate triangulate or trapezoidal, with connective sutures converging backwards except in members of the Warburgellinae where they diverge backwards; hypostome elongate or rectangular, anterior wings generally conspicuous, median body well inflated; thorax typically of 10 segments, in rare cases with 8 or 9; preannulus may or may not be present; pygidial margin entire or spinose; axis with 3–33 rings, pleural areas with 3–14 pleural ribs; pygidial border may be present; exoskeleton smooth or with sculpture of fine raised striations, granules or with a combination of both.

Discussion. This diagnosis is composed with reference to that of the Proetacea by Richter, Richter & Struve *in* Moore (1959, p. O382), and it is accepted that all the Permo-Carboniferous 'phillipsiid' trilobites are members of the Proetidae, following Hahn & Hahn (1967) and Osmólska (1970). Subfamily diagnoses are given with reference to that of the family. Many different classifications of the Proetidae have been proposed, particularly since 1945. Prior to that date, proetid trilobites were generally grouped into the Proetidae and the Phillipsiidae, distributed among a limited number of genera and not classified into subfamilies. The first attempt at a comprehensive classification of the Proetidae was by Barrande (1852, p. 437) who also (pp. 430–1) outlined the history of the genus *Proetus* up to 1850. Barrande listed 44 European species which he placed in *Proetus* and classified in the following way:

I Section. 8 segmens au thorax
II Section. 9 segmens au thorax

III Section. 10 segmens au thorax

A.
Contour du pygidium uni
- a. Test lisse
- b. Test granulé
- c. Test granulé et strié
- d. Test strié

B.
Contour du pygidium orné de pointes
- e. Test lisse
- f. Test granulé et strié
- g. Test strié

Barrande placed primary emphasis on the number of thoracic segments, further subdividing the groups of species on the type of pygidial margin and finally on surface sculpture.

The remainder of the nineteenth century saw little further attempt to classify the Proetidae, and few further proetid genera were proposed. Salter (1864, p. 2) placed *Proetus*, *Phaeton*, *Phillipsia*, *Griffithides* and *Brachymetopus* in the Proetidae, and Oehlert (1886) divided the proetid trilobites into two families, the Proetidae and the Phillipsiidae. The Richters' exhaustive studies from 1912 to 1956 contributed much to our knowledge of the post-Silurian Proetidae, and although they did not propose any new subfamilies until 1956 (the Cornuproetinae), their work laid the foundations for the developments after 1945.

Přibyl's (1946a) classification formed the basis on which further modifications were developed. He recognized five subfamilies within the Proetidae—the Proetinae, Phillipsiinae, Tropidocoryphinae, Prionopeltiinae and the Dechenellinae. Erben (1951) concentrated his efforts on the genus *Proetus*, recognizing a number of subgenera and groups of subgenera. Hupé (1953) greatly expanded Přibyl's classification, introducing nine further subfamilies and one new family, at the same time elevating the Tropidocoryphinae and Dechenellinae to family status. The classifications of Přibyl and Hupé use only a limited number of characters to define the subfamilies, and also suffer because the authors did not have the opportunity to see much of the material at first hand, or to see good illustrations. R. & E. Richter & Struve (*in* Moore 1959, pp. O382–98) attempted to rationalize Hupé's classification and they dispensed with many of his subfamilies. Their classification stands as the most widely used since 1959. They classified each subfamily on a number of characters, and illustrated nearly all the genera they diagnosed. Several of their illustrations, when compared with the type specimens of the type species of the genera represented, are very inaccurate (particularly those of *Proetidella*, *Warburgella*, *Astroproetus*, *Clypoproetus* and *Prantlia*, fig. 301, p. O396), and these have misled some subsequent workers. New illustrations of all these genera are available herein. The only major classification proposed since 1959 has been that of Pillet (1969), in which he proposed a further six subfamilies in addition to those already in existence; one of these (the Eremiproetinae) had already been proposed earlier by Alberti (1967). Pillet's additional subfamilies are not accepted here. Pillet gives drawings of all of the type species of the genera diagnosed, but many of these are inaccurate and have evidently been derived from the literature rather than from the original specimens, many of which were examined during the present study.

Post-Silurian Proetidae are not discussed in detail, as they lie beyond the scope of the present work. In this monograph, genera belonging to the following subfamilies are described and figured: Proetinae Salter, 1864, Schizoproetinae Yolkin, 1968, Cornuproetinae R. & E. Richter, 1956, Tropidocoryphinae Přibyl, 1946a, and Warburgellinae subfam. nov. The most satisfactory classification is to be derived from a combination of exoskeletal characters, and of these, the most fundamental appear to be the type of rostral plate and pygidial pleural ribs (see Text-fig. 2) and the presence or absence of the preannulus. Besides these, some or all of the following are used in subfamily diagnoses: (1) Glabellar outline and nature of glabellar furrows; (2) Presence or absence of preglabellar field and of tropidium; (3) Presence or absence of lateral occipital lobes; (4) Presence or absence of panderian opening at base of genal spine; (5) Number of thoracic segments; (6) Number of pygidial axial rings, and (7) Type of surface sculpture.

Subfamily Proetinae Salter, 1864

Nom. transl. Přibyl, 1946a (*ex* Proetidae Salter, 1864)

Genera and subgenera included. *Proetus* (*Proetus*) Steininger, 1831; *P.* (*Coniproetus*) Alberti, 1966; *P.* (*Gerastos*) Goldfuss, 1843; *P.* (*Lacunoporaspis*) Yolkin, 1966; *P.* (*Pudoproetus*) Hessler, 1963; *Ascetopeltis* Owens, 1973; *Cyphoproetus* Kegel, 1927; *Unguliproetus* Erben, 1951.

Diagnosis. Glabellar furrows typically not incised, commonly inconspicuous; occipital ring with or without well defined lobes; preglabellar field commonly short (sag.) or absent; genal spines may or may not be present; cephalic doublure with panderian opening at base of genal spine; thorax of 10 segments, with preannulus; pygidium with or without border, axis with 4–12 rings, pleural areas with 3–8 pairs of ribs, which are scalloped in exsagittal section (see Text-fig. 2), no distinct postaxial ridge; sculpture granular or smooth, striated only in *Ascetopeltis*.

Remarks. Of the genera included in this subfamily in Moore (1959, pp. O384–5), *Crassiproetus* Stumm is transferred tentatively to the Schizoproetinae, while unpublished work by the author suggests that *Isbergia* Warburg is related to *Panarchaeogonus* which is considered to be an otarionid. Pillet (1969, pp. 62–5) removed *Cyphoproetus* and *Unguliproetus* each to its own monotypic sub-family. Although each has characters which are atypical of the Proetinae (deep 1p furrows in the former, a long (sag.) preglabellar field in the latter), the overall sum of their characters is essentially proetine, and there is little reason to separate them.

Osmólska (1970, p. 12) included the Carboniferous genera *Bollandia* Reed, 1943, *Reediella* Osmólska, 1970, and doubtfully *Proetides* Walter, 1924, in the Proetinae. The morphology of the cranidia and pygidia of *Bollandia* and *Reediella* suggests that they are probably more closely linked with the Schizoproetinae than with the Proetinae. I agree with Hessler (1962) that *Proetides* belongs to the Brachymetopidae.

The earliest members of the Proetinae, as conceived here, are species of *Cyphoproetus*, which are represented in early Caradoc strata. *Proetus* itself has been found in small numbers in the Ashgill, but does not become important until the mid-Silurian.

Genus **PROETUS** Steininger, 1831

(=*Aeonia* Burmeister, 1843; *Forbesia* McCoy, 1846; *Trigonaspis*
Sandberger & Sandberger, 1850)

Type species. Originally designated by Steininger 1831, p. 335; *Calymmene concinna* Dalman, 1827, p. 234, pl. 1, figs. 5a–c; from Mulde Beds (Silurian, Wenlock Series), Djupvik, western side of Gotland, Sweden.

Diagnosis. Preglabellar field typically absent, but a very short (sag.) one may be developed; lateral glabellar furrows commonly inconspicuous; occipital ring with or without lobes; genal spine present or absent; pygidium with or without border, axis with 6–12 rings, pleural areas with 5–8 pairs of ribs. Surface smooth or granular.

Remarks. The genus *Proetus* has been widely misused in the past, commonly as a 'convenient' genus in which to place a wide range of proetid species. In more recent years, with attempts to rationalize proetid systematics the concept of *Proetus* has become more restricted, and attempts have been made to delineate species groups within the genus (e.g. by Erben 1951). Several subgenera have been proposed—e.g. *Longiproetus* Cavet & Pillet, 1958 and *Coniproetus* G. Alberti, 1966, but relations between them and the nominate subgenus *Proetus* (*Proetus*) have remained unclear. The concept of *Proetus* (*Proetus*) has been based largely on Devonian species such as *P. cuvieri* Steininger, 1831 and *P. bohemicus* Hawle & Corda, 1847 rather than on the type species *P. concinnus*, of Silurian age. Whittington & Campbell (1967, p. 456) and Campbell (1967, p. 15) have suggested that the subgenus *Proetus* (*Proetus*) might be confined to a group of mid- and late Silurian species centred on the type species, *concinnus*. They noted a range of common characters in this group, including: (a) glabellar shape and outline of muscle areas; (b) fine pitting on free cheek; (c) lack of distinct pygidial border; (d) incurving terrace lines on margin of pygidium (e.g. Pl. 3, fig. 5); (e) fine granulation on glabella, and (f) position of eye lobe. Besides the type species, Whittington & Campbell (*op. cit.*) considered the following to be probable members: *P. pluteus* Whittington & Campbell, 1967, *P. foculus* Campbell, 1967, *P. fletcheri* Salter, 1873 and *P. morinensis* Přibyl, 1946; to these may be added *P. osiliensis* Schmidt, 1894. All are of Silurian age. Of Devonian species, Whittington & Campbell (pp. 457–8) compared *P. cuvieri* from the Middle Devonian of the Eifel region of Germany with *P. concinnus*, pointing out that it differed in apparently minor features including proportions of the glabella, length of genal spine, presence of strong tubercles on some specimens and lack of incurved marginal terrace lines on the pygidium. They concluded that these characters were not of subgeneric importance. Pillet (1969), however, has recently resurrected Goldfuss' genus *Gerastos*, with *P. cuvieri* Steininger, 1831 as its type species, and within it recognized three subgenera, *Gerastos*, *Orbitoproetus* (type species *Proetus orbitatus* Barrande, 1852) and *Bohemiproetus* (type species *Proetus bohemicus* Hawle & Corda, 1847).

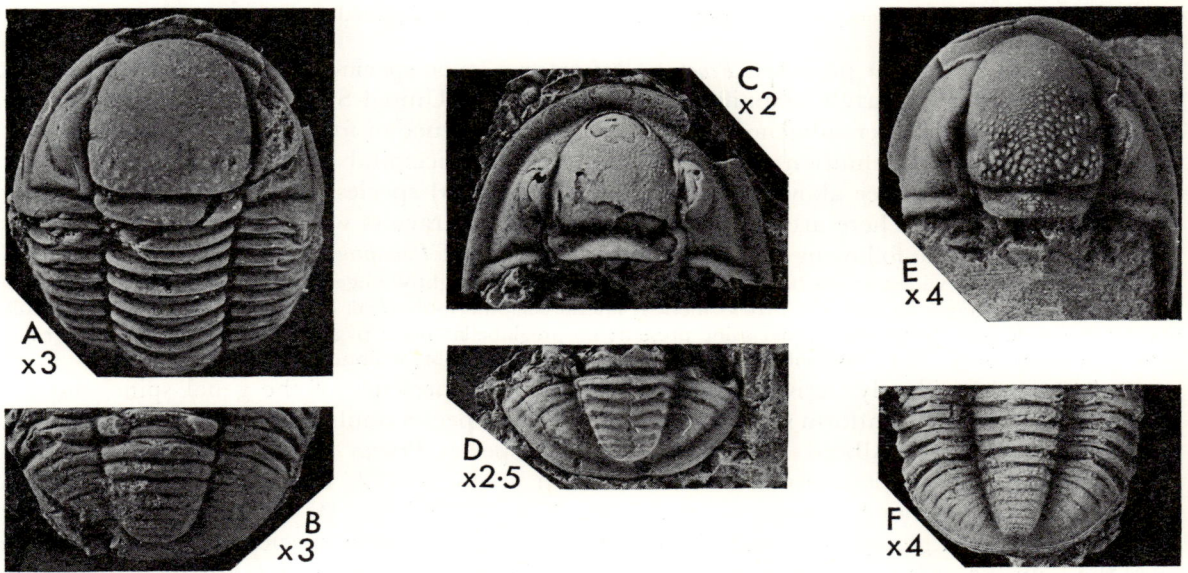

TEXT-FIG. 3. The type species of *Gerastos*, *Longiproetus* and *Coniproetus*.
   A, B. *Proetus (Gerastos) cuvieri* Steininger, 1831. Complete specimen. A, cephalon and thorax, B, pygidium. NMW 72.47G.1 (*ex* Yorkshire Museum Coll.). Middle Devonian, Gees, near Gerolstein, Eifel district, western Germany.
   C, D. *Proetus (Coniproetus) glandiferus* Novák, 1890. C, Silicone rubber cast of cephalon, NMP 813/66. D, silicone rubber cast of pygidium (on same piece of rock as cephalon). Lower Devonian, Upper Koněprusy Limestone, Zlatý kůň, near Koněprusy, Prague district, Czechoslovakia (cephalon figured by Přibyl 1965, pl. 1, fig. 7).
   E, F. *Proetus (Gerastos) tenuimargo* R. Richter, 1909 (type species of *Proetus (Longiproetus)* Cavet & Pillet, 1958). E. Cephalon, SMF 27142, F. Pygidium, SMF 27143. Both from Middle Devonian, Eifelian, Junkerburg Beds, Auburg, near Gerolstein, Eifel district.

On examining specimens of *Proetus cuvieri* (Text-figs. 3A, B) the following characters were noted which distinguish that species from members of the *Proetus concinnus* group: (a) ϵ and ξ widely separated, ϵ a wide angle (*c.* 160°); (b) lateral margin of glabella abaxially convex; (c) glabella as long as wide; (d) distinct eye platform; (e) no genal spine; (f) indistinct lateral occipital lobes; (g) no incurving terrace lines on margin of pygidium, and (h) band of parallel terrace lines around pygidial margin. Other species bearing most or all of these characters include *Proetus akrechanus* G. Alberti, 1969, *P. confragosus* Přibyl, 1965, *P. granulosus* Goldfuss, 1843, *P. microphthalmoides* Přibyl, 1965, *P. myops* Barrande, 1846, *P. orbitatus* Barrande, 1846, *P. prox* R. & E. Richter, 1956, and *P. tuberculatus* Barrande, 1846. All are of late Lower or of Middle Devonian age. The characters they all share serve to isolate this group from *Proetus* (*Proetus*), and suggest that they should be separated under the subgenus *Gerastos*, including *Orbitoproetus* Pillet.

The type species of *Proetus* (*Longiproetus*) Cavet & Pillet, 1958 is *Proetus tenuimargo* R. Richter, 1909 (Text-figs. 3E, F), and it differs essentially from *cuvieri* in its more elongated glabella, long genal spines and lack of a distinct eye platform. Species such as *Proetus dohmi* R. & E. Richter, 1918, which has been included in *Longiproetus* in the past, are almost intermediate in cephalic morphology between *cuvieri* and *tenuimargo*. *P. dohmi* could equally well be placed in *Gerastos* or in *Longiproetus*, and consequently because of the continuous gradation of species between *cuvieri* and *tenuimargo* there is no reason to retain *Longiproetus* as a separate subgenus. Incorporating *Longiproetus* with *Gerastos*, the latter may be defined as follows:—

Diagnosis. Glabella as wide (trans.), or wider than long, lateral margins abaxially convex; lateral glabellar furrows not incised; eye platform commonly present; ε and ξ widely separated, ε a wide angle (*c.* 160°); no preglabellar field; long genal spine on some species, but commonly short or absent; lateral occipital lobes indistinct or absent; pygidial border typically absent; band of parallel terrace lines around pygidial margin; sculpture granular, or exoskeleton smooth.

Hessler (1963, p. 545) proposed *Proetus* (*Pudoproetus*) (type species *Proetus fernglenensis* Weller, 1909) for a number of early Carboniferous species from the United States and Soviet Union. He distinguished it from other subgenera of *Proetus* by the presence of four pairs of lateral glabellar furrows and the poor definition or absence of the lateral occipital lobes. The latter feature is typical of *P.* (*Gerastos*) (see above), while certain *P.* (*Gerastos*) species, e.g. *P.* (*G.*) *orbitatus*, show the former character. There are, however, a number of characters which distinguish *Pudoproetus* from *Gerastos*, and the following combination is diagnostic of *Pudoproetus*:—

Diagnosis. Glabella widest across basal lobes, ranging from being slightly longer (sag.) than wide to wider than long, 1p (and commonly 2p and 3p) furrows incised; lateral occipital lobes weak or absent; eye platform typically absent; eye socle may be present; genal spine present; no preglabellar field; pygidium with border (only weakly defined in some species); 7–11 pygidial axial rings; 5–8 pairs of pleural ribs; sculpture granular.

Pudoproetus is evidently derived from *Gerastos*, and the presence of the genal spine and the commonly lacking eye platform suggest derivation from a species similar to *P.* (*Gerastos*) *tenuimargo*.

Proetus (*Coniproetus*) Alberti, 1966 has as its type species *Proetus condensus* Přibyl, 1965. The holotype of this species is a cranidium, and Alberti (1966, p. 112) diagnosed the subgenus on cranidial characters alone. Among other species, Alberti (1969, p. 93) included *Proetus finitimus* Přibyl, 1965 and *P. glandiferus* Novák, 1890 in *Coniproetus*. Both these species and *P. condensus* are from the Upper Koněprusy Limestone of the Prague district, Czechoslovakia, and on examination of silicone rubber casts of specimens of each, I can make no specific distinction between them. Hence *glandiferus* becomes the senior synonym, and is effectively the type species of *Coniproetus*. The entire cephalon and the pygidium of *P.* (*C.*) *glandiferus* are known (Text-figs. 3C, D), and exhibit the following characters: (a) weakly inflated, coniform glabella; (b) broad, flattened anterior border; (c) well-developed lateral occipital lobes; (d) long genal spine; (e) pygidium with well-developed border, and (f) band of parallel terrace lines around pygidial margin.

Proetus bohemicus, type species of *Proetus* (*Bohemiproetus*) Pillet, 1969 differs from *P.* (*Coniproetus*) *glandiferus* principally in having a more inflated glabella, a more convex anterior border, a distinct eye platform and a short genal spine. These differences cannot be considered of subgeneric importance, and *bohemicus* is here included in *P.* (*Coniproetus*), which is defined thus:—

Diagnosis. Glabella coniform; lateral glabellar furrows not incised; short (sag.) preglabellar field commonly developed; lateral occipital lobes well-defined; eye platform typically absent; long genal spines on most species; pygidial border well-developed; band of parallel terrace lines around pygidial margin; exoskeleton smooth, or with granular sculpture.

Of the species listed by Alberti (1969, p. 93) as belonging to *P.* (*Coniproetus*), *P. ryckholti* Barrande, 1846 (which lacks the well-defined lateral occipital lobes, distinct pygidial border and band of parallel terrace lines around the pygidial margin) and *P. signatus* Lindström, 1885 (here considered to belong to *Lacunoporaspis*, see below), should be excluded.

Yolkin proposed the genera *Lacunoporaspis* (type species *L. contermina* Yolkin, 1966—see Yolkin 1968, pl. 2, figs. 1–9; pl. 3, figs. 1–9), *Khalfinella* (type species *Proetus carinatus* Khalfin, 1948—see

Yolkin 1968, pl. 6, figs. 1–10) and *Ganinella* (type species *Dechenella batchatensis* Tchernysheva, 1951—see Yolkin 1968, pl. 13, figs. 1–9) based on Devonian material from SW Siberia, U.S.S.R., and assigned the first two to the Dechenellinae and the last to the Schizoproetinae. Yolkin (1968, pp. 9 and 20) distinguished *Khalfinella* from *Lacunoporaspis* by the degree of glabellar taper and by the numbers of pygidial axial rings and pleural ribs, but did not compare either with *Ganinella*. However, a comparison of his figures of each type species does not enable a generic distinction to be made between them, and all are included here in *Lacunoporaspis*, the senior synonym. The diagnostic characters of *Lacunoporaspis* (see below) are found in certain Silurian species such as *Proetus signatus* Lindström, 1885, *P. conspersus* (Angelin, 1854) and *P. obconicus* Lindström, 1885. The last named is, in some ways, transitional between *Proetus (Proetus)* and *Lacunoporaspis*, and a gradual change can be observed between the two genera in such species as this and *signatus*. Because of this evidently close phylogenetic relationship, *Lacunoporaspis* is considered here to be a subgenus of *Proetus*, and consequently to be a member of the Proetinae rather than the Dechenellinae, although it is likely to be associated with the line connecting the two subfamilies.

Five subgenera, *Proetus*, *Gerastos*, *Pudoproetus*, *Coniproetus* and *Lacunoporaspis* are recognized within the genus *Proetus*, and most Devonian and Carboniferous and a few Silurian species can be accommodated within them. There are, however, a number of (chiefly Silurian) species which cannot be assigned to any of them, and until more is known about these species and their relationships with existing subgeneric groupings, I prefer to refer to them as '*Proetus* (s.l.)'.

Species	Genal spine		Glabella		Preglabellar field		Pygidial border		Number of pygidial axial rings	Sculpture
	present	absent	tapering	parallel sided	present	absent	present	absent		
astringens	X		X			X		X	5–6	?
berwynensis	?		X		X			X	5	smooth
haverfordensis	X		X		X			X	8–9	"
latifrons		X	X			X		X	6–7	?
falcatus	X			X		X	weakly developed		6	smooth
concinnus	X		nearly		X			X	7–8	"
confossus	X		X		X		X		6	granular, pitting in cheeks
obconicus	X		X		some-times			X	7–8	smooth, pitting in cheeks

TABLE 1. Summary of diagnostic characters of *Proetus* species described herein.

Subgenus **PROETUS** Steininger, 1831

Types species. As for genus.

Diagnosis. Glabella elongate, commonly weakly constricted laterally; field of free cheek with fine pitting, generally rather inconspicuous; prominent genal spine and distinct lateral occipital lobes always present; no preglabellar field; pygidium without border and with occasional incurved marginal terrace lines; surface smooth.

Remarks. Comparative remarks are given under *Proetus* (above).

Proetus (Proetus) concinnus (Dalman, 1827)   Pl. 2, figs. 14, 15; Pl. 3, figs. 1–9

	1827	*Calymmene concinna* Dalman, p. 234, pl. 1, figs. 5a–c.
	1831	*Proetus concinnus* (Dalman); Steininger, p. 335.
non	1839	*Asaphus concinnus* (Dalman); Emmrich, p. 35.
	1839	*Calymene* indeterminable; Murchison, pl. 14, fig. 5.
	1843	*Gerastos concinnus* (Dalman); Goldfuss, p. 538.
non	1843	*Aeonia concinna* (Dalman); Burmeister, p. 40, pl. 1, fig. 5.
	1845	*Proetus concinnus* (Dalman); Lovén, p. 49, pl. 1, figs. 2a, b.
non	1846	*Proetus concinnus* (Dalman); Beyrich, pl. 3, figs. 8–10.
non	1846	*Aeonia concinna* (Dalman); Burmeister, p. 100, pl. 5, fig. 8.
	1848	*Proetus sp.*; Salter *in* Phillips & Salter, pl. 6, figs. 2–4.
	1854	*Proetus sp.*; Murchison, pl. 27, fig. 8.
	1854	*Forbesia concinna* (Dalman); Angelin, p. 22, pl. 17, fig. 5.
non	1857	*Proëtus concinnus* (Dalman); Nieszkowski, p. 556.
	1873	*Proetus Fletcheri* Salter, p. 134.
	1877	*Proetus Fletcheri* Salter; Woodward, p. 56.
	1885	*Proetus concinnus* (Dalman); Lindström, p. 78.
	1891	*Proetus Fletcheri* Salter; Woods, p. 151.
	1901	*Proetus Fletcheri* Salter; Reed, p. 11, pl. 1, figs. 5, 6.
	1901	*Proetus concinnus* (Dalman); Lindström, p. 67, pl. 6, figs. 19–23.
	1923	*Proetus concinnus* (Dalman); R. & E. Richter, p. 240.
	1946	*Proetus (Proetus) concinnus* (Dalman); Přibyl, p. 4.
	1956	*Proetus (Proetus) concinnus* (Dalman); R. & E. Richter, p. 353, pl. 5, figs. 29, 30, text-fig. 2.
	1959	*Proetus (Proetus) concinnus* (Dalman); R. & E. Richter & Struve *in* Moore, p. O385, fig. 293, 1a–c.
	1961	*Proetus sp.*; Mitchell, Pocock & Taylor, p. 30 (*partim*).
	1967	*Proetus concinnus* (Dalman); Whittington & Campbell, p. 456, pl. 3, figs. 4, 5, 9, 11, 12.
	1967	*Proetus fletcheri* Salter; Whittington & Campbell, p. 456.
	1967	*Proetus concinnus* (Dalman); Campbell, p. 11.
	1967	*Proetus fletcheri* Salter; Campbell, p. 15.
	1968	*Proetus concinnus* (Dalman); Martinsson, p. II, figs. 1A–C.
	1969	*Proetus (Proetus) concinnus* (Dalman); Alberti, p. 74.
	1969	*Proetus (Proetus) concinnus* (Dalman); Pillet, pl. 1, fig. 1; pl. 6, fig. 1.

Holotype. By monotypy; UM G733, Pl. 3, figs. 5a–c; a complete, partially enrolled specimen with external surface preserved, figured by Dalman 1827, pl. 1, figs. 5a–c; from Mulde Beds (Silurian, Wenlock Series), Djupvik, parish of Eksta, W side of Gotland, Sweden.

Material, horizons and localities. On Gotland, Sweden, *P. (P.) concinnus* is common in the Mulde Beds. In the British Isles it has been collected at many localities in the Wenlock Shale, Wenlock Limestone and Lower Elton Beds. Specimens include: from the top of the Wenlock Shale— NMW 71.6G.202, 204, 208, NMW 72.18G.30–31 from large bedding plane exposure on E side of Wren's Nest Hill, Dudley, Worcestershire (SO 9370 9180)[1]; from undifferentiated Wenlock Shale —SM A28251–53, OUM C802, GSM 36163–65 from Dudley, OUM C686, C770, C787, C800 from Malvern Tunnel; from Tickwood Beds—GSM JD2165 from old quarry 410 yd at 278° from Pilgrim Cottage, Longville in the Dale (SO 5491 9383); from Wenlock Limestone—NMW 71.6G.246 from bedding plane exposure 230–270 yd SW of 'Caves' public house, Wren's Nest Hill, Dudley, Worcestershire (SO 9350 9210); BU1821–1836, NMW 27.110G.873, SM A10248 (lectotype of *fletcheri*), SM A28263–65, SM A28268–73, BM I1513, GSM 36400 from Dudley (exact locality unspecified); NMW 71.6G.432 from old roadside quarry in Harton Hollow Wood *c.* ¾ mile S of Harton, Wenlock Edge (SO 4803 8761); NMW 71.6G.433 from road cutting on W side of Longville–Stanway road, 1¾ miles NNE of Rushbury, Wenlock Edge (SO 5397 9269); NMW 71.6G.323 from Coates Quarry, 1¾ miles SW of Much Wenlock, Wenlock Edge (SO 6045 9935); NMW 72.18G.3–7 from Hayes Quarry, 1½ miles SW of Much Wenlock, Wenlock Edge (SO 6015 9915); NMW 72.18G.18, 19 from debris near entrance to Shadwell Rock Quarry, *c.* ½ mile N of Much Wenlock (SJ 6240 0088); GSM JD2816, 2853 from old quarry 700 yd at 245° from New House Farm, Church Stretton (SO 4608 9408); GSM FGD2549, 2554–7, 2562, 2570

---

[1] Six or eight-figure numbers with two-letter prefixes are National Grid References.

from stream 300 yd at 235° from Whettleton (SO 4386 8221); GSM JD2745, 2755 from old quarry 630 yd at 81° from Eaton church (SO 5057 9007); GSM Dr906, 922 from old quarry 130–150 yd S of Fetterlocks Farm, 2360 yd ESE of church at Shelsley Beauchamp, Worcestershire (SO 752 632); NMW 71.6G.290, 292 from Clencher's Mill, $2\frac{1}{4}$ miles SE of church at Ledbury, Herefordshire (SO 732 349); NMW 71.6G.294, 298, from 'swimming pool quarry', $\frac{1}{2}$ mile N of Eastnor Castle, Herefordshire (SO 7370 3782); NMW 72.18G.60–65 from Eastnor Castle Quarry, 400 yd W of Gold Hill Farm, Herefordshire (SO 7322 3629); GSM 36393, pygidium, probably from Rock Farm, May Hill (see Stubblefield 1938, p. 32, footnote), GSM 36394–97 from 'Eastnor' (exact locality unknown); GSM GSb4049 from 'Ledbury Hill' (exact locality unknown); NMW 71.6G.159 from Hobbs, c. $\frac{1}{2}$ mile NE of Longhope, May Hill, Gloucestershire; GSM 33123–24 from 'Rock Farm, May Hill'; NMW 71.6G.420–21 from Borstal Institute Quarry, $\frac{3}{4}$ mile ESE of Common Coed-y-paen, 3 miles SW of Usk, Monmouthshire. From Lower Elton Beds and 'Lower Ludlow'— NMW 71.6G.426 from stream section, $\frac{3}{8}$ mile NNW of Upper Millichope, Wenlock Edge (SO 518 899); NMW 72.18G.1 from streamside exposure 1100 yd NE of Westhope, Wenlock Edge (SO 4781 8699); GSM 36748–49 from Cut-throat Lane, Ledbury, Herefordshire; NMW 29.62G.25–26 from Woolhope, Herefordshire.

Diagnosis. Glabella typically weakly constricted at γ and overhanging anterior border, but may be laterally bowed or pyriform and not overhanging anterior border. Pygidium with 7–8 axial rings, posterior end of axis clearly defined; pleural areas with five or six pairs of ribs, and occasional incurved marginal terrace lines.

Description. Cephalon rather strongly vaulted with broad, weakly convex border defined by deep, narrow, anterior and lateral border furrows. Glabella longer (sag.) than wide (trans.) in majority of specimens examined, and typically nearly parallel-sided, weakly constricted opposite γ and overhanging anterior border (e.g. Pl. 3, figs. 5a, c), but may be laterally bowed (e.g. Pl. 3, fig. 4) or pyriform (e.g. Pl. 3, figs. 7, 8), not overhanging anterior border. Varies from being strongly to rather weakly convex in lateral and longitudinal profiles. Three pairs of lateral glabellar furrows, indicated as darker areas on glabellar surface. 1p: opposite centre of palpebral lobe, turning backwards almost at right angles at mid-length, dying out before reaching occipital furrow. Inconspicuous auxiliary impression associated with it. 2p: nearly opposite γ, simple, gently oblique inwards and backwards. 3p: short distance in front of 2p, nearly transverse.

Occipital furrow deeper and wider than axial, medially nearly transverse, turning gently forwards in front of lateral occipital lobes. Anterior slope nearly vertical, posterior inclined at c. 45°. Occipital ring a little wider (trans.) than glabella, with distinct, ovate, inflated lateral lobes, defined by furrows which shallow rapidly abaxially. Prominent median tubercle.

Anterior branches of facial sutures weakly divergent, γ close to axial furrow. Posterior branches with ϵ and ξ as independent angles, the intervening stretch close to and parallel with axial furrow. Palpebral lobe backwardly placed, close to glabella, and about $\frac{1}{2}$ its sagittal length, steeply inclined from axial furrow, flattening abaxially. Eye large, crescentic, about $\frac{2}{3}$ sagittal length of glabella. Eye socle narrow, curb-like, lower margin not defined by incised furrow and runs parallel with upper margin.

Field of free cheek narrow, gently convex and steeply declined from eye region. Below eye it is as wide as lateral border. Posterior border furrow narrow, a little wider than lateral, running into axial furrow opposite lateral occipital lobe. Posterior border about $\frac{1}{2}$ width (exsag.) of occipital ring adaxially, but gradually widens abaxially and is about same width as latter at base of genal spine. Genal spine narrow, extending backwards as far as fifth thoracic segment. Short median groove deflected abaxially from lateral border furrow.

Cephalic doublure as wide as border, gently convex ventrally. A short distance in front of genal spine is small panderian opening (Pl. 3, fig. 9). Rostral plate trapezoidal, with posterior margin about $\frac{1}{4}$ length (trans.) of rostral suture. Hypostome with strongly convex median body, rising to a crest near anterior end. Backwardly directed median furrows shallow but distinct, dividing median body into large anterior lobe and small crescentic posterior lobe. Anterior face of anterior lobe triangular, remainder of lobe with fine terrace lines arranged in a series of concentric 'V's opening towards anterior. Lateral border rather broad, defined by deep, distinct lateral border furrow. Posterior border furrow similar to lateral, and on posterior border are two pairs of short, backwardly directed spines. Anterior border furrow deeper than lateral, anterior border narrow and ventrally upturned. Anterior wing prominent, trapezoidal and dorsally upturned, without wing process. Posterior wing rather small, triangular.

Thorax of ten segments. Axis tapers gently backwards, rather strongly convex. Intra-annular furrow shallow, not incised; on first few rings annulus and preannulus of almost equal width (sag.) but posteriorly annulus is considerably wider. Articulating furrow narrow, deep, incised, articulating half ring about $\frac{2}{3}$ the width (sag.) of annulus plus preannulus. Inner part of pleura nearly horizontal, outer part steeply declined. Narrow, incised pleural furrow runs obliquely backwards and outwards, abaxially truncated by posterior edge of articulating facet, about half way between fulcrum and end of pleura. Posterior pleural band always wider (exsag.) than anterior. Abaxially pleura bluntly truncated. Articulation between thoracic segments, cephalon and pygidium the same as that of *P. (P.) pluteus* (Whittington & Campbell 1967, p. 453, pl. 2, figs. 23–25, 31; cf. this monograph, Pl. 3, fig. 9).

Pygidium subparabolic, without border. Axis tapers gently backwards to clearly defined, bluntly rounded posterior end, consisting of seven or eight rings which become progressively more ill-defined towards posterior. First ring narrower and elevated higher than remainder, separated from them by broad, deep ring furrow. Pleural areas convex with five or six pairs of ribs with scalloped profile. Pleural and interpleural furrows commonly rather ill-defined (e.g. Pl. 3, figs. 3b, 5b), but are more distinct on some specimens (e.g. Pl. 3, fig. 2). First pair of pleural furrows deeper and more distinct than remainder, and all die out before reaching margin. Pygidial doublure weakly convex ventrally, slightly constricted behind axis.

Exoskeleton smooth apart from fine, inconspicuous pitting on free cheek and terrace lines on doublure and on lower part of cephalic border. Occasionally pygidial marginal terrace lines turn inwards onto pleural areas.

Measurements.

Cranidia	A	A_1	A_2+A_3	A_4	K	$\delta-\delta$	
UM G 773 (E)	—	5·0	—	1·0	3·8	5·8	HOLOTYPE
OUM C686	9·7	6·8	1·4	1·5	5·5	7·9	
BU 1836 (E)	8·9	6·8	1·0	1·1	7·7	5·0	
SM A10248 (E)	8·7	5·6	1·5	1·6	6·1	8·6	Lectotype of *fletcheri*
GSM 36748 (E)	(8·0)	(6·0)	(1·0)	(1·0)	(4·7)	(7·8)	
BM I1513 (E)	6·2	4·4	0·8	1·0	3·6	5·8	
GSM 33124 (E)	5·8	4·4	0·5	0·9	3·6	5·9	
BU 1821 (E)	5·5	3·9	0·8	0·8	3·4	4·9	
SM A28269 (E)	5·2	3·7	0·7	0·8	3·3	4·6	
BU 1822 (E)	4·7	3·3	0·7	0·7	2·9	4·1	
BU 1831 (E)	3·7	2·3	0·8	0·6	3·0	2·2	
SM A28271 (E)	2·7	1·8	0·5	0·4	1·3	2·0	

Pygidia	Z	Y	W	X	
UM G773 (E)	4·1	3·3	7·1	2·8	HOLOTYPE
OUM C686 (E)	6·8	5·6	12·4	5·4	
GSM GSb4049 (E)	6·5	6·3	(9·8)	3·3	
SM A10248 (E)	6·1	5·2	10·9	4·1	Lectotype of *fletcheri*
GSM 36393 (E)	5·8	5·5	7·8	3·0	
BM I1513 (E)	4·6	3·5	7·2	2·9	
BU 1821 (E)	3·2	2·7	6·8	2·1	
GSM 36749 (E)	3·9	3·5	7·8	2·4	
SM A28269 (E)	3·1	2·5	5·9	2·0	
BU 1822 (E)	2·3	1·9	5·4	2·0	
BU 1831 (E)	2·3	1·8	4·8	1·2	

Remarks. P. (P.) concinnus is one of a complex of closely related species of Wenlock and early

Ludlow age, which includes *P.* (*P.*) *pluteus* Whittington & Campbell, 1967, *P.* (*P.*) *foculus* Campbell, 1967, *P.* (*P.*) *osiliensis* Schmidt, 1894 and *P.* (*P.*) *moriensis* Přibyl, 1946.

Specimens from the Wenlock of the British Isles here considered to belong to *P.* (*P.*) *concinnus* have previously been referred to *P. latifrons* or to *P. fletcheri*, many specimens in old museum collections bearing one or other of these names. *P. fletcheri* Salter, 1873 was first described formally by Reed (1901, p. 11) [Salter (1873, p. 134) listed two numbers (a825, a828) for this species in his *Catalogue of Cambrian and Silurian fossils*. Two specimens were incorporated within a825 and one within a828 (see Reed 1901, p. 11). The most perfect of the three is one of those labelled a825, now numbered SM A10248, and was figured by Reed (1901, pl. 1, figs. 5, 6) and is probably the one upon which Salter's (1873, p. 134) original drawing was largely based. This specimen is here selected as the lectotype of *fletcheri* and figured on Pl. 3, figs. 4a, b]. Although the glabellar outline of the lectotype of *fletcheri* differs from that of the holotype of *concinnus* (cf. Pl. 3, figs. 4a, 5a), other parts of the exoskeleton are exceedingly similar. Other specimens (e.g. Pl. 3, figs. 7, 8) differ from the type specimens of both *concinnus* and *fletcheri* in having a rather weakly inflated pyriform glabella, but again the remainder of the exoskeleton is very similar. A range of glabellar shapes between these three extremes can be observed, and as there are no other exoskeletal differences, these may be considered to be the product of spatial and ecological variation, which seems to be borne out by the preservation of specimens with *fletcheri*-like and pyriform glabellar types in grey mudstones, and specimens with the *concinnus* glabellar types in purer limestones. As the majority of the material is from old, vaguely localized collections, it is difficult to assess the amount of variation to be expected in a single population.

Subgenus **LACUNOPORASPIS** Yolkin, 1966
(=*Khalfinella* Yolkin, 1968; *Ganinella* Yolkin, 1968)

Type species. Originally designated by Yolkin 1966, p. 28; *Lacunoporaspis contermina* Yolkin, 1966, p. 28, figs. 1–5; from Devonian, early Eifelian, SW Siberia, U.S.S.R.

Diagnosis. Glabella conical or pyriform, with weakly incised furrows; short (sag.) preglabellar field may be present; occipital ring commonly with lateral lobes; pygidium with narrow axis with 7–12 well defined rings; broad pleural areas with 5–8 pairs of pleural ribs of scalloped profile, pleural and interpleural furrows typically narrow and sharp and of similar depth; pygidial border well developed in all but a few early species; marginal terrace lines may be incurved on to edge of pleural areas; sculpture granular, sometimes distinct pitting developed on cheeks.

Remarks. Because *Khalfinella* and *Ganinella* are here considered congeneric with *Lacunoporaspis* (see above), Yolkin's (1966, p. 26) diagnosis of the latter is accordingly emended. It is also enlarged by including certain additional characters considered to be diagnostic.

1. **Proetus (Lacunoporaspis) confossus** sp. nov.   Pl. 4, figs. 1–8; Text-fig. 4

Name. Latin 'confossus', meaning full of holes; from the distinctive pitting on the cheeks.

Type specimens. Holotype, NMW 71.6G.502, cranidium; Pl. 4, figs. 1a–c; paratypes, NMW 72.18G.34, free cheek, Pl. 4, fig. 4, NMW 71.6G.244, pygidium, Pl. 4, fig. 7; all from Wenlock Limestone, Nodular Beds (Silurian, Wenlock Series), large bedding plane exposures 230–270 yd SW of 'Caves' public house on W side of Wren's Nest Hill, Dudley, Worcestershire (SO 9350 9210).

Material, horizon and locality. This species has only been recorded from the type locality, which in addition to the above has furnished: NMW 71.6G.241–43, NMW 71.6G.245, NMW 71.6G.500, NMW 72.18G.33, BM It8869, cranidia; NMW 71.6G.501, NMW 71.6G.504, NMW 72.18G.35, free cheeks, and NMW 71.6G.503, NMW 72.18G.36–37, pygidia.

Diagnosis. Glabella conical, approximately as long (sag.) as wide (trans.), tapering rapidly forwards, not laterally constricted; no preglabellar field; palpebral lobe and eye small; genal spine very short; pygidial axis with 6 rings; pygidial border developed, sculpture of fine granules, distinct pitting on cheeks.

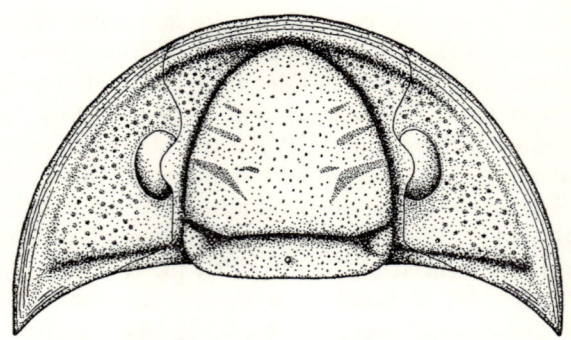

TEXT-FIG. 4. Reconstruction of the cephalon of *Proetus* (*Lacunoporaspis*) *confossus* sp. nov. (based on Pl. 4, figs. 1, 3, 4). ×6 approx.

Description. Cephalon with narrow border, more clearly defined anteriorly than laterally (cf. Pl. 4, fig. 1 and fig. 4), deep narrow anterior border furrow confluent with preglabellar furrow. Glabella conical, tapering rapidly forwards to blunt point, approximately as long (sag.) as wide (trans.), not constricted, moderately inflated. Three pairs of weakly incised glabellar furrows; 1p: abaxial end opposite centre of palpebral lobe and from this point runs backwards at about 45°, first widening, then narrowing to a point, dying out before reaching occipital furrow. Associated with 1p is an auxiliary impression (see Text-fig. 4). 2p: opposite γ, clavate, widening adaxially and backwardly directed at about 70°. 3p: not far in front of 2p, transversely elongated, isolated from axial furrow.

Occipital furrow deep, median part nearly transverse, lateral ends curving forwards, anterior slope nearly vertical, posterior slope inclined backwards at low angle. Occipital ring as wide (trans.) as glabella, and wider (sag.) than anterior border. Small, ovate, lateral lobe developed, also small median tubercle. Stretch β–γ of anterior branch of facial suture diverges at 20°–26° from an exsagittal line through γ, which is close to axial furrow. Posterior branches with ϵ and ξ widely separated, with intervening stretch running close to and parallel with axial furrow.

Palpebral lobe small, between $\frac{1}{3}$ and $\frac{1}{4}$ sagittal length of glabella. Eye small. Field of free cheek broad, posterior border furrow deeper than lateral. Genal spine short, with wide panderian opening close to its base on ventral surface (see Pl. 4, fig. 5). Cephalic doublure as wide as border, with five to six prominent terrace lines which are bunched together near margin and also where they curve round abaxial edge of panderian opening.

Thorax unknown.

Pygidium with narrow axis (the apparently broad axis of Pl. 4, fig. 6 is due to crushing) with six well defined rings. Pleural areas with four to five pleural ribs which curve gently backwards, truncated at inner edge of border. Pleural and interpleural furrows of approximately equal depth, anterior and posterior pleural bands of about same width (exsag.).

Sculpture of very fine granules on glabella, occipital ring and pygidium. Anterior portion of fixed cheek and field of free cheek with small pits.

Measurements.

Cranidia	A	A₁	A₂+A₃	A₄	K	δ–δ	
	A	A_1	A_2+A_3	A_4	K	$\delta-\delta$	
NMW 71.6G.502 (E)	6·5	4·5	0·7	1·3	4·4	5·8	HOLOTYPE
NMW 71.6G.241 (E)	7·8	6·0	0·8	1·0	(5·5)	—	
BM It8869 (E)	7·2	5·2	0·9	1·1	5·0	(6·1)	
NMW 71.6G.500 (E)	6·2	4·5	0·8	0·9	4·3	5·3	
NMW 71.6G.242 (E)	4·6	3·3	0·6	0·7	2·8	3·9	
Pygidia	Z	Y	W	X			
NMW 71.6G.503 (E)	5·1	4·1	(9·4)	3·1			
NMW 71.6G.244 (E)	4·9	4·0	7·6	2·0	PARATYPE		

Remarks. This species closely resembles *P.* (*L.*) *obconicus* Lindström from which it may be distinguished as follows: the glabella is proportionately broader and more strongly tapered forwards (cf. Pl. 4, figs. 1–3 and figs. 11–13); the palpebral lobe is smaller; there is a smaller number of pygidial axial rings (six), compared with seven to eight.

In the collections of the Institute of Geological Sciences there are two cranidia (GSM PS274–5) and a pygidium (GSM PS279) from beds of late Wenlock age at Pistyll-Dewi Quarry, 1200 yd SE of Llanarthney church, Carmarthenshire (SN 5427 1955). Proetids from this locality were listed by Thomas (*in* Strahan *et al.* 1907, p. 46) as *Proetus* n. sp. and *Proetus latifrons* McCoy but they bear a close resemblance to *Proetus confossus*, the principal differences being their more elongated and narrower glabella and narrower pygidial axis (cf. Pl. 4, figs. 1–3, 6, 7, and figs. 9, 10). The significance of the differences in the Pistyll-Dewi specimens is difficult to assess without more material, and pending further information these specimens are referred to as *P.* (*L.*) cf. *confossus*.

2. **Proetus (Lacunoporaspis) obconicus** Lindström, 1885   Pl. 4, figs. 11–19, Pl. 5, fig. 1

1848   *Proetus latifrons* M'Coy; Salter, p. 337, pl. 6, figs. 1, 1a–c.
1854   *Forbesia latifrons* M'Coy; McCoy, p. 174.
1854   *Proetus latifrons* M'Coy; Murchison, p. 235, Foss. 46, fig. 7.
1873   *Proetus latifrons* M'Coy; Salter, p. 165.
1885   *Proetus obconicus* Lindström, p. 78, pl. 15, figs. 22–24.
1904   *Proetus pseudolatifrons* Reed, p. 78 *partim, non* pl. 11, figs. 7–9.
1916   *Proetus signatus* Lindström; Reed, p. 168, pl. 8, fig. 12.
1938   *Proetus* cf. *signatus* Lindström; Stubblefield, p. 32.
1954   *Proetus* sp.; R. & E. Richter, p. 20, pl. 1, fig. 17.
1969   *Proetus*? sp.; Alberti, p. 368, pl. 46, fig. 8.
1969   *Proetus (Coniproetus) obconicus* (Lindström, 1885); Alberti, p. 446.

Type specimens. From Lindström's three syntypes, RM Ar29035, a complete enrolled specimen whose cephalon was figured by Lindström (1885, pl. 15, fig. 23), is here selected as lectotype. It is refigured on Pl. 5, figs. 1a–d. The other two specimens, RM Ar29033, a cranidium, and RM Ar29034, a pygidium, are paralectotypes. All from Eke Beds, Rhizophyllumkalken (Ludlow Series), probably Leintwardine or Whitcliffe Stage), Lau Backar, Gotland, Sweden.

Material, horizons and localities. Disarticulated remains, most commonly preserved as internal with counterpart external moulds, and occasional more complete specimens, have been found in the Upper Bringewood and Leintwardine Beds at numerous localities. *P.* (*L.*) *obconicus* reaches its acme at about the boundary of the Lower and Upper Leintwardine Beds, at which horizon it is the commonest trilobite. Material and localities include: from Upper Bringewood Beds—NMW 71.6G.434–39 from forestry track exposure on SE side of Mary Knoll Valley, 2 miles SW of Ludlow church (SO 486 728); from Lower Leintwardine Beds—Cophall Hollow, 1½ miles ENE of Leintwardine, Herefordshire (SO 4276 7471); NMW 71.6G.518–528 from old quarry in dingle S of Marlow lane, 980 yd E of Marlow Farm, near Leintwardine (SO 4087 7675); LCM 321′1970/16 from exposure in Marlow Lane (in channel fill deposits) (SO 4067 7678); trackside exposure ⅝ mile S of Lawnwell Barn, near Edwards Farm, Leintwardine (SO 422 758); BM It8822–23, NMW 71.6G.182–3, 185 from 'Goggin' lane section, 1 mile SE of Elton, Herefordshire (SO 472 701); NMW 71.6G.320 from old quarry on S side of forestry track on SE side of Mary Knoll Valley, 2 miles SW of Ludlow (SO 489 725); NMW 72.23G.2 from roadside quarry 1080 yd ENE of Mary Knoll House, near Ludlow (SO 4910 7399); BM It8821 from quarry at S end of Slang Coppice, NW of Holloway Farm, Wenlock Edge (SO 5341 8926); BM It8820 from quarry at SE end of Bache Plantation, Siefton Batch, Wenlock Edge (SO 4772 8477); exposure by track due S of Fernhall Mill, Wenlock Edge (SO 4985 8680). From Upper Leintwardine Beds—NMW 71.6G.134–5, 143, 149, 151–2 from small disused quarry in Docker Parks, 3 miles NE of Kendal, Westmorland (SD 551 954); exposure at Pont-shoni, 3 miles SE of Builth Wells, Brecknock (SO 078 468); stream section from Downton-on-the-Rock to Bow Bridge, Shropshire (SO 4292 7374); LCM 321′1970/15 from Church Hill, Leintwardine (SO 4120 7368); LCM 321′1970/3, 7, 12, 13 from Lawnwell Dingle, *c.* 1¾ miles NE of Leintwardine (SO 4167 7678–9); NMW 71.6G.302, 303,

306, 308 from exposure on N side of road between Shelderton and Shelderton Rock, Shropshire (SO 416 778); NMW 71.6G.447 from exposure on S side of 'Bengry track', on N edge of Beechenbank Wood, 300 yd NW of Aymestrey church (SO 423 654); NMW 71.6G.177, BM It589, It591 from 'Goggin' lane section, 1 mile SE of Elton (SO 472 701); NMW 71.6G.443, 4 from forestry track exposure on S edge of Haye Park Wood, *c.* 2½ miles SW of Ludlow church (SO 488 712); quarry on Diddlebury–Middlehope road, 220 yd NE of Fernhall Mill, Wenlock Edge (SO 5006 8660); GSM DEW4000, 4012 from lane section, Llandegfedd, Monmouthshire (ST 3435 9703 to 3426 9695).

In addition to the above, there are a number of vaguely localized specimens in old museum collections likely to have originated from the Leintwardine Beds which include the following: GSM 36859 (Salter 1848, pl. 6, figs. 1a–c) from Usk, above the castle; SM A16602 (Reed 1916, pl. 8, fig. 12) from Hilla Farm, 1 mile NE of Bettws Newydd, near Usk; GSM GSb4060–65, GSM 36856–8, SM A37163–71, SM A38493–95 from the Kendal and Underbarrow districts, Westmorland.

Diagnosis. Glabella elongate, conical, weakly inflated; cheek and lateral parts of preglabellar field pitted; long genal spine; prominent lateral occipital lobes; pygidium without distinct border; axis narrow with 7–8 rings; 5–6 pairs of pleural ribs; exoskeleton smooth, with sporadic, localized granules.

Description. Cephalon with rather wide, weakly convex border, defined by shallow but distinct anterior and lateral border furrows. Glabella elongate, conical, commonly not laterally constricted, but weakly so in some specimens. Three pairs of lateral glabellar furrows, scarcely impressed on surface; 1p: opposite middle of palpebral lobe, extending about half way towards sagittal line, abaxial end directed weakly backwards, adaxial end running almost exsagittally, not reaching occipital furrow. Small auxiliary impression between anterior part of 1p and sagittal line, and there appears to be a second, between distal end and occipital furrow (Pl. 4, fig. 13). 2p: a little posterior to γ, directed backwards at about same angle as abaxial part of 1p, and extending farther adaxially than 1p. 3p: a small, ovate area, about ⅓ way from 2p to anterior end of glabella.

Occipital furrow deep and rather wide with steep anterior and shallow posterior slope, narrowing laterally in front of lateral occipital lobes. Occipital ring ranges from being a little wider (sag.) to a little narrower (sag.) than anterior border, and maintains constant width laterally. Lateral occipital lobes small, ovate, incompletely isolated, fused with remainder of occipital ring laterally. Small median tubercle present.

Minute preglabellar field seen on some specimens (e.g. Pl. 4, fig. 12), but is commonly absent (Pl. 4, fig. 17). Its length is exaggerated on internal moulds (e.g. Pl. 4, fig. 11). Section β–γ of anterior branch of facial suture diverges abaxially forwards from γ at 20°–30°. Posterior branches with ϵ and ξ as independent angles, the intervening stretch running close to and parallel with axial furrow. Palpebral lobe crescentic, rising at a shallow angle from axial furrow. Eye between ½ and ⅓ sagittal length of glabella. Eye socle narrow, with non-incised lower margin running parallel with upper.

Field of free cheek gently convex. Posterior border furrow broader and deeper than lateral, truncated at base of genal spine, which is broad and long, with a shallow median groove. Cephalic doublure ventrally convex, with fine parallel terrace lines. On some internal moulds (e.g. Pl. 4, figs. 11, 16) the impression of the triangular rostral plate may be seen.

Hypostome elongated, with well vaulted median body. At its anterior end is distinct triangular facet. Abaxial ends of median furrows about ¼ way along median body from posterior, and directly behind them are well defined maculae. Anterior, lateral and posterior parts of hypostomal border narrow, defined by wide, deep border furrow. One pair of spines on posterior margin. Terrace lines on median body arranged in a 'U' shape, with base of 'U' towards posterior.

Thorax of ten segments. Axis rather narrow, tapering backwards so that last ring is about ⅔ width (trans.) of first. Preannulus about ½ width (sag.) of annulus. Pleura with broad, shallow pleural furrow which dies out abaxially where it is truncated by posterior edge of articulating facet. Anterior and posterior bands of pleura of more or less equal width (exsag.). Posterolateral corner of each pleura angular.

Pygidium subparabolic, without border. Axis at anterior end about $\frac{1}{3}$ pygidial width on external surface, about $\frac{1}{4}$ on internal moulds, with seven or eight rings. First two or three ring furrows deep, conspicuous. More posterior ones become progressively shallower. Pleural areas broad and weakly convex, with five to six pairs of pleural ribs which curve gently backwards, widening slightly adaxially, and are scalloped in longitudinal section. Pleural and interpleural furrows nearly parallel, former deeper, both dying out before reaching margin. Occasional marginal terrace lines extend inwards from margin, some running into pleural furrows (e.g. Pl. 4, fig. 14). Pygidial doublure like cephalic, with distinct parallel terrace lines.

Surface smooth, with sporadic, localized granules. Pitting on free cheek and lateral part of glabellar field.

Measurements.

Cranidia	A	A$_1$	A$_2$+A$_3$	A$_4$	K	δ–δ	
RM Ar29035 (E)	—	4·4	—	0·9	3·7	5·0	LECTOTYPE
BM It8820 (E)	12·0	8·2	2·0	1·8	7·2	9·6	
SM A16602 (I)	8·8	5·8	1·6	1·4	5·5	(7·4)	
RM Ar29033 (E)	8·5	6·3	1·2	1·0	5·3	7·1	PARALECTOTYPE
GSM 36859 (I)	6·9	5·0	1·1	0·8	4·0	(5·8)	
GSM GSb4065 (I)	(5·8)	3·8	1·4	—	3·5	(4·7)	
NMW 71.6G.522 (I)	5·0	3·2	1·0	0·8	3·0	4·8	
NMW 71.6G.181b (I)	3·7	2·5	0·6	0·6	2·3	—	

Pygidia	Z	Y	W	X	
RM Ar29035 (E)	4·0	3·0	7·8	2·5	LECTOTYPE
RM Ar29034 (E)	6·7	5·5	(10·0)	3·1	PARALECTOTYPE
BM It8821 (E)	(5·8)	(4·9)	10·4	3·5	
GSM 36859 (I)	4·9	4·0	9·5	2·4	
BM It8822 (I)	4·4	3·5	8·8	2·2	
GSM GSb4065 (I)	3·3	2·8	8·6	2·2	
NMW 71.6G.177 (I)	1·5	1·1	2·5	0·5	

Remarks. *P. (L.) obconicus* and *P. (L.) confossus* are the only *Lacunoporaspis* species occurring in the British Silurian. On Gotland three other species are found in addition to *P. (L.) obconicus*—*P. (L.) verrucosus* Lindström, *P. (L.) conspersus* (Angelin) and *P. (L.) signatus* Lindström. These species all differ from *obconicus* in having coarser, denser granular sculpture and in possessing well developed pygidial borders, the latter feature being characteristic of Devonian *Lacunoporaspis* species. In its pygidial morphology, *obconicus* is almost transitional between *Proetus (Proetus)* species and such *Lacunoporaspis* species as those mentioned above, and might be considered as being one of the most primitive members of the subgenus; it is probably a descendant of *Proetus (Proetus)* on the line leading to *Proetus (Lacunoporaspis)*.

PROETUS s.l. (subgenera incerta)

The following species are not assigned to a named subgenus. They probably belong to several new subgenera.

1. Proetus (s.l.) berwynensis (Whittington, 1966a)   Pl. 1, fig. 1

1966a *Astroproetus berwynensis* Whittington, p. 83, pl. 25, figs. 14–16.
non 1966a *Astroproetus berwynensis?* Whittington, p. 84, pl. 26, fig. 1 (? = *Decoroproetus papyraceus*).

Holotype. BM 59355, Pl. 1, fig. 1; a complete internal mould with counterpart external mould, figured Whittington 1966a, pl. 25, figs. 14–16; probably from Dolhîr Beds (Ashgill Series), Cynwyd, 2$\frac{1}{2}$ miles SW of Corwen, Merioneth. The only specimen known.

Diagnosis. Glabella coniform, with weakly impressed lateral furrows; occipital ring without lateral lobes; very short (sag.) preglabellar field apparently present; genal spine apparently absent; pygidial axis short, with 5 rings; no pygidial border; surface smooth.

Description. See Whittington 1966a, p. 83.

Remarks. Whittington (1966a, p. 83) assigned this species to *Astroproetus*, but revision of that genus (see below) shows that *P. berwynensis* has a number of characters atypical of it (e.g. lack of lateral occipital lobes, presence of the preannulus, scalloped pygidial pleural ribs). The overall characters of this species instead indicate assignment to *Proetus*, and as such it is one of the earliest known representatives. *P. berwynensis* shows closest resemblance to *P. astringens* sp. nov. (Pl. 2, figs. 7, 11, 12) and to *P. latifrons* (McCoy) (Pl. 1, figs. 11, 12; Pl. 2, figs. 1, 2), particularly in glabellar outline; although unlike *P. berwynensis*, *astringens* and *latifrons* have lateral occipital lobes.

2. Proetus (s.l.) cf. berwynensis (Whittington, 1966a) Pl. 1, figs. 2–7

1909 *Proetus* cf. *brachypygus* Marr & Nicholson; Cantrill *in* Strahan *et al.*, p. 58.

Material, localities and horizons. All the specimens referable to *Proetus* cf. *berwynensis* originate from Ashgill strata, which probably belong to the Cautleyan Stage (Dr. D. Price, personal communication, 1971) exposed in the area between Carmarthen and Haverfordwest: GSM Pr213, pygidium, from 'Bala Limestone', 230 yd E of Bron-haul Farm, 2400 yd SSE of church at St. Clear's, Carmarthenshire (SN 2843 1357); BM In54446, pygidium, from 'Robeston Wathen' Limestone, Trewern Quarry, 400 yd N of Fron, 1½ miles W of Whitland, Carmarthenshire (SN 1722 1728); numerous cranidia, pygidia and free cheeks, including BM It8851–55, NMW 72.9G. 11–12, from decalcified limestone overlying Robeston Wathen Limestone, old quarry in dingle 430 yd N of Robeston Wathen church, Pembrokeshire (SN 0843 1615).

Measurements.

Cranidium	A	A_1	A_2+A_3	A_4	K	δ–δ
BM It8854 (I)	5·4	3·8	0·7	0·9	3·8	5·0

Pygidia	Z	Y	W	X
BM In54446a (I)	3·1	2·7	5·2	2·3
BM It8853 (I)	2·7	2·1	4·6	1·9
GSM Pr213 (I)	2·6	2·0	5·3	2·0

Remarks. The *Proetus* specimens from the Ashgill of SW Wales are very similar to *P. berwynensis*, the essential difference being the presence of the genal spine in the former and its apparent absence in the latter. Further material of *P. berwynensis* may show that the absence of the genal spine in the holotype is merely due to bad preservation, and if so, there is no reason not to include the material from SW Wales in the same species.

3. Proetus (s.l.) haverfordensis sp. nov. Pl. 1, figs. 8–10

Name. From Haverfordwest, Pembrokeshire, near which the type locality is situated.

1914 *Proetus stokesi* (Murchison); Jones *in* Strahan *et al.*, pp. 108, 246.
1914 *Proetus* sp.; Jones *in* Strahan *et al.*, pp. 109, 246.

Holotype. SM A32743, Pl. 1, figs. 9a, b; a complete, distorted internal mould, with counterpart external mould; from Uzmaston Beds (Llandovery Series, late Fronian or early Telychian Stage), section below path SW of Uzmaston farm, the Frolic, near Haverfordwest, Pembrokeshire (SM 965 145).

Material, horizons and localities. Outside the type locality, this species has been recorded only from the Canaston Beds (Llandovery, Telychian Stage), represented by the following: GSM TJ655, GSM Pr2571, from small quarry by roadside at Valley Gate, ¾ mile W of Narberth Bridge (SN 0955 1430); GSM TJ779 from exposures on S bank of Eastern Cleddau, 1000 yd SW of Blackpool Bridge (SN 0527 1398).

Diagnosis. Glabella conical; genal spine apparently short: pygidium large, without border, a little longer (sag.) than thorax: pygidial axis about 70% of pygidial length (sag.) with 8–9 rings; pleural areas broad, with only first two pairs of ribs distinct; marginal terrace lines incurved; exoskeleton smooth.

Description. Cephala of available specimens are badly preserved, but show the following characters: cephalic border rather broad, defined by shallow cephalic border furrow (the apparently deep lateral border furrow seen on the left hand side of the holotype (Pl. 1, fig. 9b) is exaggerated by crushing). Glabella conical, rather weakly inflated, wider (trans.) than long (sag.). All specimens too badly preserved to show lateral glabellar furrows. Occipital ring with ovate lateral lobes. Short (sag.) preglabellar field apparently present (e.g. Pl. 1, fig. 8). Section β–γ of anterior branch of facial suture diverges abaxially forwards from γ at about 30°, palpebral lobe crescentic, eye approximately $\frac{1}{2}$ sagittal length of glabella. Genal spine seems to be short. Impression of triangular rostral plate can be seen on holotype internal mould (Pl. 1, fig. 9b).

Thorax with ten segments. Axis narrow, at no point wider (trans.) than pleurae. Pleural furrow narrow, incised, truncated by posterior edge of articulating facet and dividing pleura into wider posterior band and narrower anterior band. Posterolateral corner of pleura rounded.

Pygidium large, subparabolic, without border and slightly longer (sag.) than thorax. Axis rather narrow, about 70% of total pygidial length and consisting of eight to nine rings, which become indistinct at posterior end. Pleural areas broad, gently convex. First two pairs only of pleural ribs are distinct, the rest being hardly apparent. Pleural and interpleural furrows parallel, former narrow and sharp (first two or three pairs), latter broad and shallow. Towards anterior, marginal terrace lines of pygidium incurved, running subparallel with pleural and interpleural furrows for short distance. Pygidial doublure broad, gently convex ventrally with prominent terrace lines running parallel with margin, inner ones being bunched together behind axis.

Exoskeleton, as far as can be judged, is smooth.

Measurements.

Cranidium	A	A_1	A_2+A_3	A_4	K	δ–δ	
SM A32743a (I)	—	(3·7)	(1·2)	—	4·8	(5·3)	HOLOTYPE

Pygidia	Z	Y	W	X		
SM A32743a (I)	7·2	5·1	12·4	3·3	HOLOTYPE	
GSM TJ779 (I)	5·4	3·7	9·0	2·3		

Remarks. This species is distinctive in having a large pygidium, which is proportionately larger than in most *Proetus* species (compare, for example, *P.* (*P.*) *concinnus*, Pl. 3, figs, 3,5), and invites comparison with the more or less contemporaneous *Crassiproetus*? *curtisi* (Pl. 7, fig. 7). The relationship between these two species is difficult to gauge with the poor material available, but the markedly conical glabella distinguishes *P. haverfordensis* from *C.*? *curtisi* and from *Crassiproetus*. The incurved marginal terrace lines on the pygidium are a feature typical of *Proetus* (*Proetus*).

4. **Proetus** (s.l.) **latifrons** (McCoy, 1846)   Pl. 1, figs. 11, 12; Pl. 2, figs. 1, 2, 4
    1846 *Forbesia latifrons* McCoy, p. 49, pl. 4, fig. 11.
non 1848 *Proetus latifrons* (M'Coy); Salter, p. 337, pl. 6, figs. 1, 1a.
non 1854 *Forbesia latifrons* M'Coy; McCoy, p. 174.
non 1854 *Proetus latifrons* (M'Coy); Murchison, p. 235, Fossils 46, fig. 7.
non 1857 *Proëtus latifrons* M'Coy sp.; Nieszkowski, p. 558.
non 1873 *Proetus latifrons* (M'Coy); Salter, p. 165.
non 1879 *Proetus latifrons* (M'Coy); Nicholson & Etheridge, p. 171.
non 1904 *Proetus latifrons* (M'Coy); Reed, p. 76, pl. 11, figs. 4, 4a.
non 1907 *Proetus latifrons* (M'Coy); Thomas *in* Strahan *et al.*, p. 46.

Type specimens. In the collections of the National Museum of Ireland, Dublin, there are two specimens of *Proetus latifrons* in the Griffith Collection, both of which would have been available to McCoy. One of these, NMI.G8:1970, a complete internal mould, is almost certainly the specimen upon which McCoy's pl. 4, fig. 11 was based, and is here selected as the lectotype. The other specimen, NMI.G9:1970, an external mould of an incomplete thorax and pygidium is the paralectotype. Both specimens from Silurian, supposed Upper Llandovery beds at Egool, Ballaghaderreen, Co. Roscommon, Ireland.

Material, horizons and localities. Apart from the type material, there are a few specimens (NMW 71.6G.104–7; NMW 71.6G.495–497) from Wenlock shale, small exposures along path of old mineral railway, *c.* 400 yd SE of Moon's Hill Quarry, near Stoke St. Michael, Mendip Hills, Somerset (ST 6647 4561).

Diagnosis. Glabella broadly conical, wider (trans.) than long (sag.); three pairs of weak glabellar furrows; posterior branch of facial suture with ϵ and ξ as independent angles; no genal spine; pygidium without border.

Description. Cephalon roughly semicircular, border rather narrow, weakly convex and defined by shallow but distinct anterior and lateral border furrows. Glabella broadly conical, wider (trans.) than long (sag.), weakly convex in longitudinal profile, more strongly so in lateral profile. It tapers rapidly forwards to broad, rounded frontal lobe. Three pairs of lateral glabellar furrows; 1p opposite centre of palpebral lobe, directed backwards at about 35°. 2p: opposite γ, shorter than 1p, not so strongly backwardly directed. 3p: short distance in front of 2p, directed backwards at about same angle and isolated from axial furrow. No preglabellar field.

Occipital furrow more or less transverse running weakly forwards abaxially. Occipital ring badly preserved on all available material, but is apprently a little wider (sag.) than anterior border and not quite as wide (trans.) as greatest width of glabella. Large, ill-defined lateral occipital lobes seem to be present. Section β–γ of anterior branch of facial suture diverges abaxially forwards at 14°–20° from γ. Posterior branches with ϵ and ξ as independent angles, the intervening stretch close to and parallel with axial furrow. Palpebral lobe subparabolic, a little under $\frac{1}{2}$ sagittal length of glabella. Eye crescentic, half glabellar length. Field of free cheek moderately convex, quite steeply declined from eye. Posterior border furrow apparently wider and deeper than lateral. Genal angle rounded, no genal spine.

Cephalic doublure weakly convex ventrally with terrace lines running parallel to margin. On lectotype, mould of trapezoidal rostral plate seen.

Thorax of ten segments. Axis tapers backwards so that last ring is about 75% of width of first. Anteriorly axis wider than pleurae, but posteriorly is narrower. Each pleura with narrow, distinct pleural furrow, truncated near abaxial end.

Pygidium subparabolic, without border. Axis with six to seven clearly defined rings, strongly arched in longitudinal profile. Pleural areas with three to four pairs of ribs, with shallow, rather ill-defined pleural and interpleural furrows, neither reaching margin. Pygidial doublure like cephalic, with similar parallel terrace lines.

No specimens are sufficiently well preserved to show nature of surface sculpture.

Measurements.

Cranidia	A	A_1	A_2+A_3	A_4	K	δ–δ	
NMI.G8:1970 (I)	(6·5)	5·0	0·9	(0·6)	5·9	(7·0)	LECTOTYPE
NMW 71.6G.496 (I)	6·8	4·6	1·0	1·2	5·0	(6·4)	
NMW 71.6G.495a (I)	—	3·2	0·7	—	3·8	—	
Pygidia	Z	Y	W	X			
NMI.G8:1970 (I)	(4·5)	—	9·9	3·7	LECTOTYPE		

The synonymy list for this species indicates the extent to which it has been misinterpreted, following the original misinterpretation of Salter (1848, p. 337). Reed (1904, pp. 78–79) attempted to rectify the state of affairs, and his conclusions are discussed under *Astroproetus pseudolatifrons* (Reed) (see p. 60).

Proetus latifrons as interpreted here is restricted to the probable Upper Llandovery of NW Ireland and the Wenlock shale of the Mendip Hills. Most museum specimens from Dudley and Malvern bearing the name *Proetus latifrons* belong to *Proetus (Proetus) concinnus* (Dalman). *Proetus latifrons* is distinctive in its glabellar outline and lack of genal spines, characters shared by *Proetus granulatus* Lindström (Pl. 2, fig. 3) from the Silurian of Gotland. Because of poor preservation, it is unknown whether the coarse granular sculpture of the latter is also present in the former. Better material of *P. latifrons* is required to assess its affinities more fully.

5. **Proetus** (s.l.) cf. **latifrons** (McCoy, 1846)   Pl. 2, figs. 5, 6, 10

1958 *Proetus* sp.; Curtis, p. 141, pl. 29, figs. 3a, b.

Material, horizons and localities. Known only from the Llandovery Series: NMW 72.40G.1, pygidium, from Coralliferous Series (Telychian Stage, highest C_6) cliff exposure 400 yd SW of Little Marloes Farm, Marloes Bay, Pembrokeshire (SM 7873 0718); SM A39560, almost complete, distorted internal mould from Coralliferous Series (C_6), 80 ft above base, Marloes Bay; GSM 35997, GSM 90031, pygidia from Damery Beds (Telychian Stage, C_5), Damery Quarry, near Tortworth, Gloucestershire (ST 7045 9440) [ZCM loc. 82]; OUM C1363, cranidium, from Huntley Hill Beds (Fronian Stage, C_{1-2}), old quarries in Newent Wood, 400 yd at 315° from Folly Farm, SW side of May Hill, Gloucestershire (SO 7014 2104) [ZCM loc. 4]; OUM C1926 cranidium, from Huntley Hill Beds (Fronian or Telychian Stage, C_{3-4}), trackside exposures immediately W of old quarry on E side of Nottswood Hill, 760 yd at 242° from Hinders Farm, May Hill (SO 7047 1811) [ZCM loc. 11]; OUM C16117, cranidium, from Pentamerus Beds, 830 yd W. of Marshbrook, Shropshire (SO 4341 8982) [ZCM loc. 61].

Description. The specimens listed above are characterized by coniform glabella, occipital ring with well developed lobes, no preglabellar field, no(?) genal spine, strongly tapering pygidial axis with seven rings, at least four pairs of pleural ribs, and no pygidial border.

Measurements.

Cranidia	A	A_1	A_2+A_3	A_4	K	δ–δ
OUM C1363a (I)	—	4·3	—	0·8	4·7	5·8
OUM C1926b (I)	(6·0)	4·1	1·0	(0·9)	3·8	—
OUM C16117a (I)	4·8	3·1	0·9	0·8	2·9	—

Pygidia	Z	Y	W	X
GSM 35997 (I)	5·5	4·6	(7·4)	2·7
GSM 90031 (I)	4·4	3·8	7·1	2·1

Remarks. These specimens of Llandovery age are all ill-preserved, but as far as can be seen their affinities lie with *P. latifrons*. The glabellar shape is broadly similar, and the apparent absence of a genal spine might be taken as a second similar character. They differ essentially in the proportionately narrower, more pointed glabella and appear to have a narrower pygidial axis, but better preserved material of *P. latifrons* may well show them to be conspecific. The major difference from the contemporaneous *P. haverfordensis* is seen in the pygidial axis—short in that species and long in *P.* cf. *latifrons* (cf. Pl. 1, figs. 9, 10 and Pl. 2, figs. 5, 10). Cranidia and pygidia (e.g. NMW 72.40G. 2–4) from probable late Wenlock Coralliferous Series of Upper Winsle, south Pembrokeshire (SM 8333 0929) are also comparable with *P. latifrons*, but their preservation is too poor to ascertain whether they are conspecific with it or with the form described as *P.* cf. *latifrons*.

6. **Proetus** (s.l.) **falcatus** sp. nov.   Pl. 2, figs. 8, 13

Name. From Latin 'falcatus', meaning curved like a sickle; referring to the path of the anterior branch of the facial suture.

Holotype. BM 59022, Pl. 2, figs. 13a–c; almost complete specimen with external surface preserved; from Lower Ludlow (Silurian), Dudley, Worcestershire.

Material, horizons and localities. OUM C772–75, complete specimens with external surface preserved from Wenlock Shale, Malvern Tunnel; BU 1819, complete enrolled specimen from Wenlock Limestone, Dudley.

Diagnosis. Cephalic border broad, well-defined; glabella parallel-sided, elongate, weakly inflated with 2 pairs of weakly impressed furrows, lateral occipital lobes poorly defined; anterior branch of facial suture strongly abaxially convex; pygidial axis with 6 rings and 5 pairs of pleural ribs; ill-defined pygidial border; exoskeleton smooth.

Description. Cephalon weakly vaulted with broad, well-defined border. Glabella parallel-sided, elongate and rounded anteriorly, defined by narrow conjoined axial and preglabellar furrows, the latter anteriorly running into anterior border furrow. No preglabellar field. Glabella weakly convex in longitudinal and transverse sections. Two pairs of weakly impressed glabellar furrows; 1p: opposite anterior part of palpebral lobe, directed obliquely backwards at about 45°, dying out shortly before reaching occipital furrow. 2p: opposite γ, directed obliquely backwards at about 80°, extending about $\frac{2}{3}$ of way towards sagittal line.

Occipital furrow deep, nearly transverse, deflected weakly forwards abaxially. Occipital ring $\frac{2}{3}$ sagittal width of anterior border, transversely a little wider than glabella, in lateral profile gently inclined towards posterior. Lateral occipital lobes triangulate, poorly defined. Small median tubercle present.

Section β–γ of anterior branch of facial suture diverges abaxially forwards at 31°–35° from γ, which is close to axial furrow. Posterior branch with ϵ and ξ as independent angles, separated only by a short distance, intervening part of suture close to and parallel with axial furrow.

Palpebral lobe subparabolic, a little under half glabellar length, posteriorly placed, and in transverse section slopes up at 35° from axial furrow, flattening abaxially. Margin at δ elevated almost to height of sagittal region of glabella. Eye large, crescentic, about $\frac{1}{2}$ glabellar length and surmounting a narrow, curb-like eye socle. Field of free cheek narrow, convex. Posterior border narrower than lateral, posterior border furrow narrower, but as deep as lateral. Genal spine extends as far back as eighth thoracic segment, median groove short, only apparent at anterior end.

Thorax of ten segments. Axis tapering backwards only slightly. Annulus markedly longer (sag.) than preannulus, intra-annular furrow weak. Annulus weakly convex in lateral profile. Inner part of pleura horizontal, but abaxially from fulcrum pleura quite steeply declined. Pleural furrow narrow and incised, truncated abaxially by broad articulating facet not far from fulcrum. Posterior pleural band somewhat wider than anterior. Distal end of pleura bluntly truncated.

Pygidium sub-semicircular, nearly twice as wide as long. Axis about 30% of total pygidial width anteriorly, tapering gently backwards to bluntly rounded posterior end. Axial furrows distinct, not incised, shallowing distinctly towards posterior. Six axial rings, defined by shallow ring furrows and becoming progressively less distinct towards posterior. Five pairs of pleural ribs, first two pairs of pleural and first pair of interpleural furrows reaching margin, rest truncated by poorly developed border. Pleural furrows narrow and incised, interpleural shallow and inconspicuous.

Exoskeleton smooth, with exception of inner portion of genal spine, which has longitudinal terrace lines.

Measurements.

Cranidia	A	A₁	A₂+A₃	A₄	K	δ–δ	
BM 59022 (E)	4·6	3·0	1·0	0·6	2·3	(3·6)	HOLOTYPE
BU 1819 (E)	7·1	4·9	1·3	0·9	3·3	4·9	
OUM C775 (E)	7·0	4·7	1·6	0·7	3·2	5·7	
OUM C774 (E)	(6·7)	4·6	(1·1)	1·0	4·0	6·7	
OUM C772 (E)	5·9	4·3	0·9	0·7	3·4	5·1	
OUM C773 (E)	5·3	3·3	1·2	0·8	2·4	4·0	

Pygidia	Z	Y	W	X	
BM 59022 (E)	2·9	2·2	5·1	1·8	HOLOTYPE
OUM C774 (E)	4·3	3·5	7·9	2·8	
OUM C775 (E)	4·0	3·0	7·0	2·3	
OUM C773 (E/I)	(3·0)	(2·5)	5·0	1·3	

Remarks. Proetus falcatus is particularly distinctive in the path of the anterior branch of the facial suture, the narrow elongate glabella and in the weakly developed pygidial border. The glabellar outline and the weakly developed occipital lobes invite comparison with *Proetus ryckholti* Barrande (see Pl. 2, fig. 9) from the Kopanina Beds (Ludlow) of the Prague district, Czechoslovakia, but this species does not have so wide an anterior border, has a minute preglabellar field and does not have a pygidial border. *Proetus falcatus* does not appear to be closely related to members of the *Proetus (Proetus) concinnus* group.

7. **Proetus** (s.l.) **astringens** sp. nov.   Pl. 2, figs. 7, 11, 12

Name. From Latin 'astringens', meaning contracting; refers to shape of glabella.

Holotype. NMW 71.6G.494, Pl. 2, fig. 12; a nearly complete internal mould; from Upper Elton Beds (Silurian, Ludlow Series), exposure in ditch on S side of Worcester–Hereford road (A 4103), near Storridge, Herefordshire (SO 743 480).

Material, horizons and localities. This species has been recorded in small numbers from the Upper Elton Beds and Lower Bringewood Beds in the Welsh Borderland. From the Upper Elton Beds: NMW 71.6G.279, cranidium, NMW 71.6G.280, incomplete thorax and pygidium, from the type locality; BM It8850, partially enrolled specimen from stream section 625 yd E of Upper Milli-chope, Wenlock Edge, Shropshire (SO 5255 8890). From the Lower Bringewood Beds: NMW 73.7G.1, almost complete specimen, NMW 71.6G.163, pygidium, from upper level of Woodbury Quarry, *c.* 300 yd SW of Woodbury Old Farm, 1 mile NE of Shelsley Beauchamp, Worcestershire (SO 742 638); BM It8825, cranidium with one attached free cheek from well (13 ft below surface), near quarry at W margin of Big Wood, 300 yd due S of Whitbach Farm, Wenlock Edge, Shrop-shire (SO 5306 8962); NMW 71.6G.442, pygidium from forestry track exposure on SW side of Mary Knoll Valley, 2 miles SW of Ludlow church (SO 486 728).

Diagnosis. Glabella pyriform, laterally constricted at about $\frac{2}{3}$ of its length from anterior, wider (trans.) than long (sag.), rapidly tapering forwards; no preglabellar field; free cheek with genal spine; pygidium without border, axis with 5–6 rings.

Description. Cephalon with moderately broad, weakly convex border defined by shallow but distinct anterior and lateral border furrows, former confluent with preglabellar furrow in front of glabella. Glabella pyriform, wider (trans.) than long (sag.), and weakly inflated. On the holo-type (Pl. 2, fig. 12) traces of two pairs of glabellar furrows can be seen; 1p: at broadest part of glabella, opposite centre of palpebral lobe, running gently backwards. 2p: opposite γ, running backwards at about same angle as 1p. No preglabellar field.

Occipital furrow bowed weakly forwards sagittally. Occipital ring as wide (trans.) as glabella with distinct lateral occipital lobes. Section β–γ of anterior branch of facial suture diverges abaxi-ally forwards at 26°–34° from γ, which is close to axial furrow, posterior branches with ϵ and ξ as independent angles, the intervening stretch parallel with axial furrow. Palpebral lobe approxi-mately $\frac{1}{2}$ sagittal length of glabella, eye without apparent eye socle. Field of free cheek rather narrow, posterior border furrow narrower and a little deeper than lateral. Genal spine broad-based, extending backwards as far as fifth thoracic segment, narrow median furrow offset abaxially from lateral border furrow.

Thorax of ten segments, axis tapers gently backwards, wider (trans.) than pleural areas anter-iorly, as wide posteriorly. None of the material is sufficiently well preserved to show nature and proportions of preannulus. Posterolateral corners of pleura bluntly pointed.

Pygidium sub-semicircular, without border. Axis tapers fairly rapidly backwards and is composed of five to six rings, whose defining ring furrows become progressively shallower towards posterior. Pleural areas broad with four or five pairs of ribs, but are not well preserved on any of the specimens seen.

Surface seems to be smooth.

Measurements.

Cranidia	A	A_1	A_2+A_3	A_4	K	δ–δ	
NMW 71.6G.494 (I)	3·8	2·7	0·5	0·6	2·8	3·6	HOLOTYPE
NMW 73.7G.1 (E/I)	—	3·6	—	0·7	4·2	5·3	
BM It8850 (I)	—	2·8	0·8	—	4·0	—	

Pygidia	Z	Y	W	X		
NMW 71.6G.494 (I)	2·3	1·8	(4·3)	1·7	HOLOTYPE	
NMW 73.7G.1 (E/I)	4·0	(3·2)	8·7	2·9		

Remarks. Although none of the material of *P. astringens* is particularly well preserved, it is sufficiently diagnostic to indicate the presence of a distinct species. It differs from *P.* (*P.*) *concinnus* in its glabellar outline (strongly tapered forwards rather than parallel-sided or weakly tapered forwards) and proportionately broader pygidium, and from *P.* (*L.*) *obconicus* in glabellar outline, larger palpebral lobe and eye, and shorter pygidial axis with a smaller number of rings.

Genus **ASCETOPELTIS** Owens, 1973

Type species. Original designation; *Ascetopeltis bockeliei* Owens, 1973, p. 125, figs. 1M, N; figs. 2A–K; from Tretaspis Series (Ordovician, Harju Series, Stage 5a), Holmenskjaeret, Holmen, Asker, Norway.

Diagnosis. Preglabellar field, when present, is very short (sag.); occipital ring may or may not narrow abaxially; small, ill-defined lateral occipital lobes; pygidium broadly triangulate, without border; broad bluntly terminating axis with 4–5 rings defined by very shallow ring furrows; pleural areas with 3–4 pairs of pleural ribs; surface smooth, or with sculpture of fine striations, commonly interspersed with sporadic granules.

Remarks. For discussion and comparison, see Owens (1973, p. 126).

Ascetopeltis barkingensis sp. nov.   Pl. 5, figs. 2–5

Type specimens. Holotype; SM A43029, Pl. 5, figs. 3a–d; cranidium with external surface preserved, paratypes, SM A43031, cranidium and SM A43030, pygidium; all from Ordovician, undifferentiated Ashgill, 'Barking, Dent', NW Yorkshire.

Material, horizon and locality. This species has been recorded only from the type locality, from which come, in addition to the type specimens: SM A43032–33, SM A43035, cranidia, and SM A43034, pygidium.

Diagnosis. Short (sag.) preglabellar field, about $\frac{1}{2}$ length (sag.) of anterior border; small palpebral lobe; posterior branches of facial sutures with ϵ and ξ as independent angles; occipital ring not narrowed laterally; sculpture of fine striations, not interspersed with granules.

Description. Glabella broadly conical, tapering gently forwards with well rounded frontal lobe and very slightly constricted opposite anterior part of palpebral lobe. It is widest at posterior end, and weakly inflated in longitudinal and transverse sections. Three pairs of non-incised lateral glabellar furrows which interrupt striated sculpture; 1p: abaxial end opposite anterior part of palpebral lobe, from which furrow runs backwards for about $\frac{1}{3}$ of way towards sagittal line, then turning abruptly backwards and tapering to a point, dying out a short distance in front of occipital furrow. Inconspicuous auxiliary impression associated with 1p. 2p: approximately opposite γ, nearly transverse or running weakly backwards. 3p: very inconspicuous, a short distance in front of 2p and isolated from axial furrow.

Occipital furrow deep, with shallow posterior slope and steep anterior slope; median stretch arched weakly forwards sagittally, lateral ends curved fairly strongly forwards. Occipital ring as wide (trans.) as glabella, approximately as wide (sag.) as preglabellar area, and not narrowing laterally. Weakly developed lateral occipital lobes present, and a small, distinct median tubercle. Small preglabellar field, approximately $\frac{1}{3}$ width (sag.) of anterior border. Anterior border furrow narrow and distinct, anterior border rather broad and convex.

Section β–γ of anterior branch of facial suture diverges abaxially forwards at 22°–28° from γ, which is close to axial furrow. Posterior branches with ϵ and ξ as independent angles, the intervening stretch running close to and parallel with axial furrow. Palpebral lobe small, about $\frac{1}{3}$ length of glabella.

Free cheek and thorax unknown.

Pygidium about $\frac{5}{8}$ as long (sag.) as wide (trans.), without border. Axis broad, about $\frac{3}{8}$ of total pygidial width at anterior end, composed of six rings and a short terminal piece, defined by shallow ring furrows which become progressively shallower towards posterior. On lateral ends of each ring is small depression (Pl. 5, fig. 4). Pleural areas with four, possibly five pairs of ribs which curve gently backwards, widening slightly abaxially. Pleural furrows, except first which is narrow and incised, are shallow, while interpleural furrows are shallower and only apparent at their abaxial ends. Anterior and posterior pleural bands of approximately equal width (exsag.).

Striated sculpture covers entire cranidium and pygidium. Prominent, parallel terrace lines on outer part of anterior border.

Measurements.

Cranidia	A	A_1	A_2+A_3	A_4	K	$\delta-\delta$	
SM A43029 (E)	8·6	5·5	1·5	1·6	5·9	7·2	HOLOTYPE
SM A43035 (E/I)	10·0	6·5	1·7	1·8	(6·1)	—	
SM A43033 (E)	—	6·1	—	—	6·1	—	
SM A43031 (E/I)	8·5	5·8	1·3	1·4	6·2	(7·5)	PARATYPE
Pygidia	Z	Y	W	X			
SM A43034 (E)	5·8	5·1	—	(3·5)			
SM A43030 (E)	5·8	5·0	8·8	3·5	PARATYPE		

Remarks. The characters listed in the diagnosis serve to distinguish *A. barkingensis* from other *Ascetopeltis* species. The presence of the preglabellar field, the small palpebral lobe, ϵ and ξ as independent angles and the occipital ring not narrowed laterally distinguish it from *A. bockeliei* (Owens 1973, p. 126, figs. 1M, N, figs. 2A–K), and all these characters except the first from *A. lepta* (Owens 1973, p. 130, figs. 3E–G, J). *A.* sp. 1 (Owens 1973, p. 132, fig. 6L) differs in having the surface sculpture of coarser striations and in not having ϵ and ξ as independent angles.

The three species of *Ascetopeltis* compared with *A. barkingensis* are all from the late Ordovician of Scandinavia. The small number of specimens of *A. barkingensis* are labelled as having come from 'Middle Bala, Barking, Dent'. Dr. J. K. Ingham (personal communication, 1970) has seen the specimens, but knows of no beds in the Ordovician inliers of the Cautley district of similar lithology to that in which they are preserved, nor has he any further records of the species. No Ordovician outcrops are known at Barking, so possibly the specimens are derived from erratics. The foreign occurrences of *Ascetopeltis* suggest that *A. barkingensis* is likely to have originated from beds of late Ashgill age.

Genus **CYPHOPROETUS** Kegel, 1927

Type species. Subsequently designated by Přibyl 1946a, p. 15; *Cyphaspis depressa* Barrande, 1846, p. 60; from Liteň Formation (Wenlock Series), Lištice, near Beroun, Prague district, Czechoslovakia.

Diagnosis. Glabella with deep 1p furrows, defining prominent 1p lobes; occipital ring with lateral lobes, median tubercle commonly forwardly placed on ring; very short (sag.) preglabellar field may be present; lateral margin of cephalon commonly incurved at base of genal spine; thorax of 9 or 10 segments; pygidium without border, axis with 5–8 rings, pleural areas with 3–6 pairs of ribs with scalloped profile.

Remarks. Pillet (1969, p. 72) proposed the monotypic subfamily Cyphoproetinae for *Cyphoproetus*, but the only major character of the genus not found in other proetines is the prominent 1p lobe. Other characters, particularly the preannulus and the scalloped profile of the pygidial pleural ribs, are typically proetinae. Thus there is little reason to separate *Cyphoproetus* from the Proetinae.

Many authors (e.g. Barrande 1852, p. 493; Kegel 1927, p. 636) have remarked on the striking similarity between the type species of *Cyphoproetus* and species of *Cyphaspis* (= *Otarion*), the 'otarionid' characters of the former being the prominent basal glabellar lobes and the incurving of the lateral margin of the cephalon. Other characters, (e.g. position of eyes, number of thoracic segments, presence of preannulus, structure and size of pygidium), however, demonstrate unequivocally that *Cyphoproetus* is a proetid rather than an otarionid, but is possibly derived from the latter group.

While *Cyphoproetus* shows some resemblance to otarionids, there is an even greater resemblance to warburgellines, with which it might easily be confused. There are, however, many points of difference, shown on the accompanying table, and, despite the close superficial resemblance between *Cyphoproetus* and *Warburgella*, there appears to be no close relationship, the similarities probably being due to parallel evolution.

TABLE 2. Characters distinguishing *Cyphoproetus* from *Warburgella*

	Cyphoproetus	*Warburgella*
Tropidium	Absent	Present
Transverse preglabellar ridge	Absent	Commonly present
Preglabellar field	May be present	Always present
Rostral plate	Connective sutures converge backwards	Connective sutures diverge backwards
Preannulus	Present	Absent
Pygidial pleural ribs	Scalloped profile	Flat-topped profile
Lateral margin of cephalon at base of genal spine	Commonly incurved	Never incurved

Species	Anterior border		Preglabellar field	Lateral cephalic margin at base of genal spine	Number of thoracic segments	Number of pygidial axial rings
	sagittally widened	not sagittally widened				
depressus	X		minute	incurved	10	5
facetus		X	absent	not incurved	?	4(+?)
rotundatus		X	"	"	9	4(+?)
binodosus	X		"	"	10	?5
externus		X	minute	weakly incurved on some specimens	10	4(+?)
strabismus	X		as long as anterior border	weakly incurved	?	8

TABLE 3. Summary of diagnostic characters of *Cyphoproetus* species described herein.

1. **Cyphoproetus depressus** (Barrande, 1846)   Pl. 5, figs. 6–9

1846   *Cyphaspis depressa* Barrande, p. 60.
1847   *Lichas simplex* Hawle & Corda, p. 143.
1852   *Cyphaspis depressa* Barrande; Barrande, p. 492, pl. 16, figs. 38–40.
1927   *Proetus (Cyphoproetus* n. subgen.) *depressus* (Barrande); Kegel, p. 636, pl. 32, fig. 11.
1946a   *Cyphoproetus depressus* (Barrande); Přibyl, p. 36, pl. 1, figs. 9, 9a.
1959   *P. (Cyphoproetus) depressus* (Barrande); R. & E. Richter & Struve *in* Moore, p. O386, fig. 293, 3.
1969   *Cyphoproetus depressus* (Barrande); Pillet, p. 81, pl. 3, fig. 4, p. 84, pl. 6, fig. 20.
1970   *Cyphoproetus depressus* (Barrande); Horný & Bastl, p. 117, pl. 10, fig. 6.

Type specimens. Lectotype, selected Přibyl *in* Horný & Bastl 1970, p. 117; NMP IT220; figured Barrande 1852, pl. 16, figs. 38–40, refigured Horný & Bastl 1970, pl. 10, fig. 6; from higher part of Liteň Formation (Wenlock), Lištice, near Beroun, Prague district, Czechoslovakia. A paralectotype, NMP IT221 is figured herein on Pl. 5, figs. 9a, b; horizon and locality as for lectotype.

Material, horizons and localities. This species occurs in small numbers in the Liteň Formation, Prague district, Czechoslovakia, and in the Silurian Ostracodenkalk of Lindener Mark, near Giessen, western Germany. It is represented by a number of complete or nearly complete specimens from the Wenlock Series of the British Isles: OUM C628, OUM C782–85, OUM C789–93, GSM 36140 from Wenlock Shale, Malvern Tunnel and the Wych, Malvern Hills, Herefordshire; GSM 36383, GSM 36402 from Wenlock Limestone, "Malvern"; BU 1837, GSM 36392 from Wenlock Limestone, Dudley, Worcestershire; NMW G.461 and NMW 16.116G.10, ill-preserved internal moulds, from late Wenlock, Roath Park, Cardiff.

Diagnosis. Minute preglabellar field; median occipital tubercle anteriorly placed; anterior border sagittally flattened and widened; cephalic margin incurved at base of genal spine which is narrow and lacks median furrow; thorax of 10 segments; pygidial axis with 5 rings, pleural areas with 4 pairs of ribs; sculpture granular.

Description. Cephalon moderately vaulted with distinct, convex border defined by narrow, distinct anterior and lateral border furrows. Anterior border widened and flattened sagittally. Margin distinctly incurved at base of genal spine. Glabella as long (sag.) or a little longer than wide (trans.), tapering gently forwards to bluntly rounded frontal lobe and rather weakly inflated. 1p: deep and wide at mid-length, shallowing shortly before reaching axial furrow and for a longer distance before reaching occipital furrow, into which it runs. Anterior end opposite anterior end of palpebral lobe, and from it the furrow runs backwards at about 40° to an exsagittal line, defining prominent basal lobe, which is ovate, exsagittally elongated and with independent convexity from remainder of glabella. Exsagittally basal lobe $\frac{2}{5}$ sagittal length of glabella, transversely $\frac{1}{5}$ its basal width. 2p: just anterior to γ, narrow, weakly impressed, running slightly forwards. 3p; very weak and inconspicuous, a short distance in front of 2p.

Occipital furrow with transverse median section, deepening laterally, curving first backwards, then forwards round posterior end of 1p lobes. Occipital ring about as wide (sag.) as anterior border, and wider (trans.) than glabella, nearly flat in lateral profile. Median tubercle forwardly placed, lateral lobes small, ovate. Preglabellar field minute, about $\frac{1}{4}$ sagittal width of anterior border.

Section β–γ of anterior branch of facial suture diverges abaxially forwards from γ at 20°–21°, α–β–γ describing on abaxially convex curve. γ close to axial furrow. Posterior branches with ϵ and ξ as a single angle. Palpebral lobe subparabolic, its margin elevated almost to height of sagittal region of glabella. Eye about $\frac{2}{5}$ sagittal length of glabella. Eye socle narrow, indistinct; lower marginal furrow not incised, diverging slightly from upper at each end. Field of free cheek convex. Posterior border furrow deeper and wider than lateral, posterior border a little narrower, hardly widening abaxially. Genal spine narrow based, slender, without median furrow and extending backwards as far as fourth or fifth thoracic segment.

Thorax of ten segments. Last axial ring about 70% width (trans.) of first. Preannulus about $\frac{1}{3}$ width (sag.) of annulus on anterior segments, rather less on posterior ones. Intra-annular furrow not incised, curving gently forwards abaxially to meet articulating furrow a short distance before it runs into axial furrow. Articulating half ring at least $\frac{1}{2}$ width (sag.) of annulus. Adaxial part of pleura horizontal and transverse, abaxial part beyond fulcrum is declined and directed weakly backwards. Pleural furrow narrow and incised, extending close to abaxial end of pleura, dividing it into wider posterior and narrower anterior pleural bands. Posterolateral corner of pleura rounded.

Pygidium subtriangular, a little over twice as wide (trans.) as long (sag.). Anteriorly axis about $\frac{1}{3}$ total pygidial width, tapering backwards to bluntly rounded posterior end. Five axial rings defined by shallow ring furrows, except for first which is deep. No postaxial ridge. Pleural areas gently convex, with four pairs of ribs which curve gently backwards abaxially. Pleural and interpleural furrows of similar depth, and both extend close to margin. Anterior and posterior pleural bands of similar width and convexity.

Sculpture of coarse granules.

Measurements.

Cranidia	A	A₁	A₂+A₃	A₄	K	δ–δ
BU 1837 (E)	6·1	3·8	1·2	1·1	3·3	—
OUM C784 (E)	6·0	3·7	1·4	0·9	(3·7)	—
OUM C782 (E)	4·9	3·1	1·0	0·8	(3·0)	—
OUM C791 (E)	4·6	2·6	1·2	0·8	—	—
OUM C783 (E)	3·5	2·2	0·7	0·6	2·0	2·7

Pygidia	Z	Y	W	X
BU 1837 (E)	3·2	2·4	5·2	2·1
OUM C784 (E)	2·5	2·1	6·1	2·1
OUM C791 (E)	2·0	1·8	4·6	1·5
OUM C783 (E)	1·4	1·2	2·9	1·1

Remarks. Cyphoproetus punctillosus (Lindström, 1885, p. 77) from the Wenlock of Gotland closely resembles *C. depressus*, and from Lindström's figure it appears to differ in the longer genal spine, angular posterolateral angles of the thoracic pleurae and in the proportionately longer pygidium. Other *Cyphoproetus* species are compared with *C. depressus* in Table 3.

2. **Cyphoproetus facetus** Tripp, 1954   Pl. 5, figs. 10–12; Pl. 6, fig. 1

   1954   *Cyphoproetus? facetus* Tripp, p. 671, pl. 3, figs. 13–20.
?1954   Hypostome B; Tripp, p. 686, pl. 3, fig. 9.

Type specimens. Holotype, HM A3903; internal mould of pygidium, figured Tripp 1954, pl. 3, figs. 13a–c. Paratype, HM A3904, cephalon with five attached thoracic segments, figured Tripp 1954, pl. 3, fig. 14, refigured herein Pl. 5, figs. 10a–c; both from Craighead (Kiln) Mudstones (Caradoc Series), Craighead Quarry, near Girvan, Ayrshire (NX 235 014).

Material. This species has only been recorded from the type locality, and in addition to the type specimens, the material includes HM A3905–8, BM In52679, cranidia, BM In52680, free cheek, HM A3909 and BM In52681, pygidia.

Diagnosis. Anterior border not sagittally widened; no preglabellar field; no lateral occipital lobes; lateral margin of cephalon not incurved at base of genal spine; pygidial axis with four or more rings; sculpture of fine striations and sporadic granules.

Description. Cephalon with narrow, convex border defined by narrow, distinct lateral and anterior border furrows, the latter confluent with preglabellar furrow in front of glabella. Glabella moderately inflated, tapering gently forwards, weakly constricted opposite γ and bluntly rounded in anterior. Two pairs of lateral glabellar furrows. 1p: deepest at anterior end, shallowing backwards, directed backwards at 45° and isolated from both axial and occipital furrows. Resultant ovate, partially isolated 1p lobe between $\frac{1}{2}$ and $\frac{2}{5}$ sagittal length of glabella. 2p: runs into axial furrow a short distance in front of 1p, opposite γ, is short and runs transversely or slightly forwards.

Occipital furrow deep, transverse medially, turning forwards laterally after curving backwards round posterior end of 1p lobe. Occipital ring as wide (trans.) as glabella and wider (sag.) than preglabellar area. No lateral lobes. Position of occipital tubercle unknown.

No preglabellar field. Anterior branches of facial sutures weakly divergent, γ close to axial furrow. Posterior branches with ϵ and ξ as independent angles, intervening stretch of suture close to and parallel with axial furrow. Palpebral lobe large, about $\frac{1}{2}$ sagittal length of glabella, and crescentic. Eye with narrow eye socle whose non-incised lower margin diverges from upper at either end. Field of free cheek rather narrow. Posterior border furrow narrower and deeper than lateral. Lateral margin not incurved at base of genal spine.

Cephalic doublure rather narrow.

Total number of thoracic segments unknown; most complete specimen has five preserved, which have a broad axis which hardly tapers backwards, and wider (trans.) than pleural areas. Pleurae badly preserved.

Pygidium approximately twice as wide (trans.) as long (sag.), without border. Axis with at least four rings defined by rather shallow ring furrows. Pleural areas with four(?) pairs of ribs which curve gently backwards and widen a little abaxially. Pleural furrows a little deeper than interpleural, both reach close to margin. Anterior and posterior pleural bands of approximately equal width (exsag.).

Sculpture of fine striations with sporadic granules.

Measurements.

Cranidia	A	A$_1$	A$_2$+A$_3$	A$_4$	K	δ–δ
HM A3904 (I)	3·0	2·4	0·3	0·4	(2·7)	(3·9)
HM A3906a (E)	2·2	1·5	0·2	0·5	1·6	2·3

Pygidium	Z	Y	W	X
BM In52681 (I)	0·9	0·7	1·7	0·5

Remarks. Cyphoproetus facetus is the earliest known representative of the genus. Tripp (1954,

p. 672) has pointed out the similarities between it and the Ashgill species *C. bellus* (Cooper & Kindle) from Percé, Quebec, and the Llandovery species *C. externus* Reed (see below and Table 3). Also similar is *C. rotundatus* (Begg) from the Ashgill of the Girvan district (described below).

Tripp figured a hypostome which he considered to belong to *C. facetus*, and figured a further hypostome as "Hypostome B" (1954, pl. 3, figs. 9, 19); the latter is not unlike the former, and is certainly proetid, and it is possible that both these belong to *C. facetus*.

3. **Cyphoproetus rotundatus** (Begg, 1939)   Pl. 6, figs. 2–4

1939 *Warburgella rotundata* Begg, p. 378, pl. 6, figs. 4–6.
1967 ?*Warburgella rotundata* (Begg); Ormiston, p. 62.

Type specimens. Holotype, HM A1084, Pl. 6, figs. 2a–d; a complete internal mould lacking the left free cheek and palpebral lobe, figured Begg 1939, pl. 6, figs. 5, 6; from Upper Drummuck Group (Ashgill Series, Rawtheyan Stage), Starfish Bed no. 2, Lady Burn, Girvan. Begg (p. 379) designated another complete internal mould, bearing his catalogue number BG 6458 (subsequently renumbered HM A1085) as paratype. In Begg's plate explanation the figure captions are confused, and the specimen labelled as 'paratype' (his fig. 5) is a reconstruction of the holotype. A third figure (his fig. 4), although not labelled as such, is the specimen designated as paratype by Begg and is refigured herein as Pl. 6, figs. 3a, b.

Material, horizons and localities. Begg (p. 379) stated that he had seven specimens of this species, all from the type locality. At least one of these (HM A1088) does not belong to *C. rotundatus* as it has a preglabellar field and lacks 1p lobes, and it is probably referable to *Decoroproetus asellus* (Esmark). Another specimen (HM A1087) is too badly preserved for precise identification. In addition to Begg's specimens are: BM I16056, complete internal mould with counterpart external mould and BM In43063, small internal mould, from Upper Drummuck Group, Starfish Bed no. 1, Thraive Glen, Girvan, Ayrshire.

Diagnosis. Glabella bell-shaped, distinctly constricted opposite γ; occipital ring without lobes; no preglabellar field; anterior border not widened sagittally; anterior branches of facial sutures nearly parallel; lateral margin not incurved at base of genal spine; thorax of 9 segments.

Description. Cephalon apparently with rather narrow border, but this is ill preserved on available material. Glabella bell-shaped, moderately inflated, distinctly constricted opposite γ, bluntly rounded in anterior. Only one pair of lateral glabellar furrows seen, 1p, which is deepest at mid-length, shallowing at either end where it runs into axial and occipital furrows, partially isolating prominent subovate 1p lobe which has independent convexity from remainder of glabella.

Occipital furrow deep, arched weakly forwards sagittally and turning more strongly forwards laterally after curving gently backwards behind 1p lobe. Occipital ring approximately as wide (sag.) as preglabellar area and as wide (trans.) as glabella. No lateral occipital lobes. Position of occipital tubercle unknown.

No preglabellar field. Anterior branches of facial sutures nearly parallel, with γ close to axial furrow. Posterior branches badly preserved on all specimens, but ϵ and ξ appear to be separate angles. Palpebral lobe large, almost semicircular, elevated nearly to height of sagittal region of glabella. Eye about $\frac{1}{2}$ sagittal length of glabella. Eye socle with non-incised lower margin which diverges from upper margin at either end. Field of free cheek rather narrow, declined steeply from eye region. Posterior border furrow deeper and narrower than lateral, truncated at base of genal spine which extends backwards at least as far as fourth thoracic segment. Cephalic doublure narrow, ventrally convex.

Thorax of nine segments, with axis tapering gently backwards, wider (trans.) than pleural areas anteriorly, as wide posteriorly. On the holotype (Pl. 6, figs. 2a–d) the preannulus can be seen, and is approximately $\frac{2}{3}$ the length (sag.) of annulus. Intra-annular furrow not incised, running into articulating furrow a short distance before it meets axial furrow. Pleura with deep pleural furrow which has a steep anterior and shallow posterior slope and runs close to abaxial extremity of pleura which is apparently angular. Anterior and posterior pleural bands of approximately equal width (exsag.).

Pygidium subparabolic, without border. Axis anteriorly about $\frac{1}{3}$ total pygidial width, tapers gently backwards and composed of at least four rings defined by shallow ring-furrows, but all material seen is badly preserved. Pleural areas with four pairs of ribs which curve gently backwards, widening slightly as they do so. Pleural and interpleural furrows of approximately equal depth, anterior and posterior pleural bands of about the same width (exsag.).

Measurements.

Cranidia	A	A_1	A_2+A_3	A_4	K	$\delta-\delta$	
HM A1084 (I)	3·7	2·6	0·5	0·7	(3·3)	(4·0)	HOLOTYPE
HM A1085 (I)	(2·8)	1·9	0·6	(0·3)	3·0	(3·8)	PARATYPE
BM In43063 (I)	1·7	1·2	0·3	0·2	1·5	2·3	

Pygidia	Z	Y	W	X	
HM A1084 (I)	2·6	2·2	4·8	1·7	HOLOTYPE
HM A1085 (I)	2·0	1·7	4·8	1·6	PARATYPE

Remarks. This species differs from the earlier *C. facetus* in its glabellar outline, in its more ovate 1p lobe and in the parallel anterior branches of the facial sutures, but as far as comparison is possible at present, is otherwise very similar. Comparison with other species is shown in Table 3.

Begg (1939, p. 380) was of the opinion that the lack of the preglabellar field on the type specimens of this species was due to longitudinal compression, and that normally it was present. Although the type specimens are somewhat distorted, there is no suggestion that a preglabellar field was ever present, and one paratype (HM A1088) possessing it does not belong to this species (see above).

4. **Cyphoproetus externus** (Reed, 1935)   Pl. 6, figs. 6–8

1904   *Proetus stokesi* (Murchison); Reed, p. 79, pl. 11, figs. 10, 11.
1935   *Proetus (Cyphoproetus) externus* Reed, p. 42, pl. 2, fig. 15.

Holotype. By monotypy; HM A1032, Pl. 6, figs. 6a, b; internal mould of cranidium, figured Reed 1935, pl. 2, fig. 15; from Saugh Hill Sandstones (Llandovery Series, Idwian Stage), Newlands, Girvan, Ayrshire (NS 277 044).

Material, horizons and localities. From the type locality: BM In21955, complete internal mould, BM In43162–3, BM It9089, BM It9117, cranidia; BM In21953, BM In42683, partially complete specimens from Mulloch Hill Group (Llandovery Series, Rhuddanian Stage), Mulloch Hill, near Girvan.

Diagnosis. Anterior border not widened sagittally; very short (sag.) preglabellar field present; median occipital tubercle centrally placed; lateral cephalic border weakly incurved on some specimens; thorax of 10 segments.

Description. Cephalon with rather narrow, convex border, defined by apparently shallow, narrow anterior and lateral border furrows. Anterior border not widened sagittally. Lateral border weakly incurved on some specimens (e.g. Pl. 6, fig. 8). Glabella weakly inflated, constricted opposite γ, a little longer (sag.) than wide (trans.), bluntly rounded in anterior. 1p: deepest at mid-length, shallowing at either end where it runs into axial and occipital furrows. It is directed backwards between 35° and 40°, defining elongate 1p lobe, on holotype $\frac{1}{3}$ sagittal length of glabella and a little under $\frac{1}{4}$ of its basal width. 2p: short, narrow, shallow, transverse, opposite γ, at point of greatest constriction of glabella. 3p: very inconspicuous, a short distance in front of 2p.

Occipital furrow with narrow, shallow, transverse median section, lateral ends curving first backwards, then forwards round posterior ends of 1p lobes. Occipital ring transversely somewhat wider than glabella, sagittally as wide or a little wider than anterior border. Lateral occipital lobes small, poorly developed, median tubercle situated in central position.

Preglabellar field short (sag.) about equal in width to anterior border on holotype, an internal mould, but not seen on some specimens (e.g. Pl. 6, fig. 8). Section $\beta-\gamma$ of anterior branch of facial suture diverges abaxially forwards at 20°–25° from γ, which is close to axial furrow. Posterior branches with ϵ and ζ as single angle, close to axial furrow. Palpebral lobe narrow, crescentic, inclined steeply from axial furrow, flattening abaxially. Eye approximately $\frac{1}{2}$ sagittal length of glabella, crescentic. Eye socle indistinct. Field of free cheek narrow, weakly convex. Posterior border furrow narrow, deep, abruptly truncated at base of genal spine. Genal spine extending backwards as far as sixth thoracic segment. Cephalic doublure narrow, ventrally convex. Rostral plate apparently trapezoidal.

Thorax of ten segments, axis tapering so that last ring is approximately $\frac{2}{3}$ width (trans.) of first. At anterior end axis as wide (trans.) as pleurae, but posteriorly latter wider. Pleura with distinct oblique pleural furrow, truncated abruptly close to abaxial end, dividing it into anterior and posterior bands of more or less equal width (exsag.). Posterolateral corner of pleura pointed.

Pygidium about 70% as long (sag.) as wide (trans.). Anteriorly axis occupies about $\frac{1}{3}$ total width, tapering gently backwards to bluntly rounded posterior end, and strongly convex in transverse section, consisting of at least four rings separated by shallow ring furrows. Pleural areas gently convex with three, possibly four pairs of ribs which curve gently backwards, widening slightly abaxially. Pleural furrows slightly deeper than interpleural, and both extend close to margin, running more or less parallel. Anterior and posterior pleural bands of nearly equal width (exsag.). Pygidial doublure weakly convex ventrally.

Measurements.

Cranidia	A	A_1	A_2+A_3	A_4	K	$\delta-\delta$	
HM A1032 (I)	4·6	3·0	0·7	0·9	2·6	(3·3)	HOLOTYPE
BM In21955 (I)	(4·0)	3·2	(0·4)	0·4	(3·6)	—	
BM In21953 (I)	—	3·1	—	(0·3)	2·8	3·1	
GSM 70721 (I)	3·9	2·5	0·7	0·7	2·3	—	

Pygidium	Z	Y	W	X
BM In21955 (I)	2·9	2·1	5·1	1·4

Remarks. When Reed (1935, p. 42) described *C. externus* he based it on a single cranidium, stating that it was closely allied to *Proetus* (=*Warburgella*) *stokesii*, in which he had previously (1904, p. 79) placed specimens from the Mulloch Hill Group and Saugh Hill Sandstones, but did not later transfer to *C. externus*. All the '*W. stokesii*' specimens from the Llandovery of the Girvan district have, on examination, turned out to be referable to *C. externus*.

Cyphoproetus externus most closely resembles an undescribed *Cyphoproetus* species from the Llandovery (Stage 6) of the Oslo district, but detailed comparison must await description of the Norwegian material. Comparison with other British species is shown in Table 3.

5. Cyphoproetus binodosus (Whittard, 1938)   Pl. 6, figs. 9–12

1878   *Proetus nasiger* Edgell MS; Huxley, Newton & Etheridge, p. 72 [*nom. nud.*].
1928   *Proetus* sp. undescr.; Whittard, pp. 751, 752.
1938   *Warburgella stokesi* (Murchison) [*partim*]; Whittard, p. 95, pl. 3, fig. 3, *non* fig. 1 [= *W. (W.) stokesii*], *nec* figs. 2, 2a [= *W. (W.)* sp. 1].
1938   *Warburgella binodosa* Whittard, p. 97, pl. 3, fig. 4.
1967   *Warburgella binodosa* Whittard; Ormiston, p. 62.

Holotype. GSM 36000, Pl. 6, fig. 12; internal mould of cranidium, figured Whittard 1938, pl. 3, fig. 4; from Hughley Shales (Llandovery Series, Fronian Stage, C_{2-3}), Onny River section, near Cheney Longville, Shropshire (SO 426 852).

Material, horizons and localities. From the type locality: GSM 35998–9, GSM 36001, complete or partially complete internal moulds. BM In37624, an incomplete external mould, labelled 'Upper Ordovician, Caradocian, Onny section near Cheney Longville, about $1\frac{1}{2}$ miles from Craven Arms Station' is almost certain to have originated from the Upper Llandovery, and the matrix is the same as that of specimens from the type locality.

TCD 9609, cranidium, and TCD 9608, free cheek, from Wenlock Shale, Buttington Brickworks Quarry, $2\frac{3}{4}$ miles NE of Welshpool, Montgomeryshire (SJ 264 099) are the only specimens known from outside the type locality.

Diagnosis. No preglabellar field; median occipital tubercle anteriorly placed; anterior border sagittally flattened and widened; cephalic margin not incurved at base of genal spine which is broad-based with short median furrow; thorax of 10 segments; pygidial axis with 5? rings, pleural areas with 4? pairs of ribs; sculpture granular.

Description. The available material of this species is all badly preserved, so only a rather brief description is possible. Cephalon moderately vaulted with border well defined by deep, narrow anterior and lateral border furrows, anterior border distinctly widened sagittally. Lateral border rather narrow. Glabella tapers weakly forwards, bluntly rounded anteriorly. Deep 1p furrow, which shallows at either end, running into axial and occipital furrows, defines ovate 1p lobe which has independent convexity from remainder of glabella. 2p runs into axial furrow close to 1p, and runs weakly forwards. 3p not seen.

Occipital furrow narrow and deep, median section nearly transverse, lateral ends curve round posterior ends of 1p lobes. Occipital ring as wide or somewhat wider (trans.) than glabella, and distinctly narrower (sag.) than preglabellar area. Distinct, ovate lateral occipital lobes; median tubercle forwardly placed.

No preglabellar field. Section β–γ of anterior branch of facial suture diverges abaxially forwards from γ at 13°–19°, posterior branches with ϵ and ξ as single angle. Palpebral lobe not preserved on any of specimens seen, eye large, about $\frac{1}{2}$ sagittal length of glabella. Field of free cheek gently convex. Genal spine relatively broad with short median groove, lateral margin not incurved at its base.

Thorax of ten segments, axis hardly tapering backwards. Pleura with pleural furrows dividing it into anterior and posterior bands of approximately equal width (exsag.). Posterolateral corner of pleura bluntly angular.

All pygidia very poorly preserved. Axis with five? rings, anteriorly about 28% of pygidial width and tapering gently backwards. Pleural areas with four? pairs of ribs which curve gently backwards. No border.

Sculpture granular.

Measurements.

Cranidia	A	A$_1$	A$_2$+A$_3$	A$_4$	K	δ–δ	
GSM 36000 (I)	—	4·1	2·0	—	4·7	—	HOLOTYPE
TCD 9609 (E/I)	—	4·3	—	0·9	3·8	—	
GSM 35999 (I)	5·2	3·3	1·1	0·8	3·1	(3·9)	

Pygidium	Z	Y	W	X
GSM 35999 (I)	2·9	2·3	7·0	1·9

Remarks. Whittard (1938, p. 97) considered three complete or nearly complete internal moulds bearing Edgell's manuscript name *Proetus nasiger* [*nom. nud.*] from the Hughley Shales of the Onny River section, to belong to *Warburgella stokesii* (Murchison). The shape of the rostral plate, with the connective sutures converging backwards, the lack of a preglabellar field and tropidium indicate that these specimens do not belong to *Warburgella*, but to *Cyphoproetus*. Whittard (1938, p. 97) also described a cranidium from the same locality as *Warburgella binodosa*, but the sagittally widened anterior border, and lack of tropidium show that this too belongs to *Cyphoproetus*. There is nothing to separate the Llandovery specimens described by Whittard as *W. stokesii* from this species, and they are consequently all placed here in *Cyphoproetus binodosus*.

C. binodosus is very similar to *C. depressus*, but differs because the lateral margin of the cephalon is not incurved at the base of the genal spine, it has a stouter genal spine with a short median groove, it lacks a preglabellar field, and has a less granulose sculpture.

6. **Cyphoproetus** aff. **binodosus** (Whittard, 1938)   Pl. 6, fig. 5

Material. GSM BAH1130, almost complete, partially exfoliated cephalon with parts of six thoracic segments from Hughley Shales (Silurian, Llandovery Series), Church Stretton No. 2 Borehole (Robury Ring) at depth 116 ft 2 in, 1940 yd at 343° from Methodist chapel, Asterton, Shropshire (SO 3914 9303).

Measurements.

Cranidium	A	A₁	A₂+A₃	A₄	K	δ–δ
GSM BAH1130 (E/I)	6·1	3·3	2·0	0·8	(3·0)	(3·8)

Remarks. This specimen is similar to *C. binodosus* in its broad anterior border, shape of glabella and wide (trans.) occipital ring. It differs in possessing a very wide lateral cephalic border, a short (sag.) preglabellar field, and a tropidium. The latter is typical of *Warburgella*, and is not known to occur in any species of *Cyphoproetus*, but the nature of the anterior border, the transversely wide occipital ring, the forward position of the occipital tubercle (inconspicuous on the specimen) and the apparent presence of the preannulus on the thoracic segments are all characters of *Cyphoproetus*, not *Warburgella*. Unfortunately the diagnostic rostral plate cannot be seen on the specimen. Pending further information, the specimen is included in *Cyphoproetus* as *C.* aff. *binodosus*, the species it most closely resembles.

7. **Cyphoproetus strabismus** sp. nov. Pl. 6, figs. 13–15; Pl. 7, figs. 1, 2; Text-fig. 5

Name. From Latin 'strabismus', meaning squinting; alluding to the position of the eye.

Type specimens. Holotype, TCD 9619, Pl. 6, figs. 15a–d; a partially exfoliated cephalon; paratypes, TCD 9622, cranidium; TCD 9623, incomplete thorax and pygidium; TCD 9624, pygidium; all from Trewern Brook Mudstone Formation (Wenlock Series, *lundgreni* Zone), stream near Court House, Kingswood, SW end of Long Mountain, Montgomeryshire (SJ 2463 0265).

Material. This species has only been recorded from the type locality, and in addition to the type specimens there is a number of detached exoskeletal remains: TCD 9621, 9625–28.

Diagnosis. Preglabellar field as wide (sag.) as anterior border; median occipital tubercle anteriorly placed; anterior border slightly widened sagittally; cephalic margin very weakly incurved at base of genal spine, which is narrow-based and lacks a median groove; pygidium with narrow axis with 8 rings; broad pleural areas with 6 pairs of ribs; sculpture of minute granules.

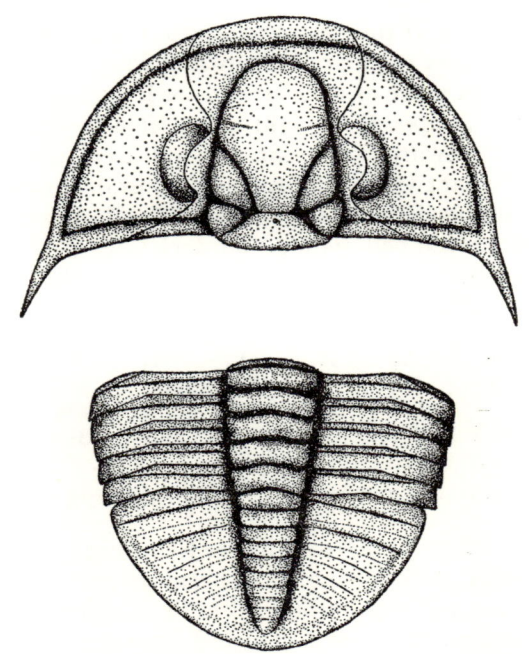

TEXT-FIG. 5. Reconstruction of *Cyphoproetus strabismus* sp. nov. (based on Pl. 6, figs. 13, 15 and Pl. 7, figs. 1, 2). ×4 approx.

Description. Cephalon with narrow border which widens slightly sagittally and which is weakly incurved at base of genal spine. Anterior and lateral border furrows narrow and well defined. Glabella weakly inflated, longer (sag.) than wide (trans.), tapering weakly forwards to bluntly rounded frontal lobe. Three pairs of lateral glabellar furrows; 1p: abaxial end opposite anterior part of palpebral lobe, and from here 1p directed obliquely backwards at about 30°, deepest at mid-length and shallowing at either end where it runs into axial and occipital furrows. Resultant elongated ovate 1p lobe between $\frac{1}{2}$ and $\frac{1}{3}$ sagittal length of glabella, from which it has independent convexity. 2p: shallow, non-incised, directed slightly obliquely backwards and joining axial furrow opposite γ, a short distance in front of 1p. 3p: about as far in front of 2p as latter is of 1p. Similar to 2p and shorter, directed slightly forwards.

Occipital furrow describes a curve which is weakly convex towards posterior, deepening at each end between 1p and lateral occipital lobes. Occipital ring narrower (sag.) than preglabellar area, and as wide or a little wider (trans.) than glabella. Prominent lateral occipital lobes and minute, forward-placed median occipital tubercle present. Preglabellar field weakly convex, as wide (sag.) as anterior border. Section β–γ of anterior branch of facial suture diverges abaxially forwards at 14°–22° from γ, which is close to axial furrow. Posterior branches with ϵ and ξ as single angle, and stretch $\epsilon + \xi$ to ω defining small triangular posterior portion of fixed cheek.

Palpebral lobe slightly over $\frac{1}{3}$ sagittal length of glabella, crescentic in outline and in transverse section inclined at about 30° from axial furrow. Eye a little over $\frac{1}{2}$ sagittal length of glabella, without distinct eye socle. Field of free cheek broad, convex, gently declined from eye region. Posterior border furrow deeper and broader than lateral. Genal spine narrow-based, rather short and without median furrow.

Where the anterior border of the holotype (Pl. 6, fig. 15a) has been damaged, the mould of the trapezoidal rostral plate can be seen. Cephalic doublure convex ventrally, with fine, parallel terrace lines.

The most complete thorax has parts of six segments preserved. Axis narrow, rather strongly convex in transverse section, and is considerably narrower (trans.) than pleural areas. Pleura with narrow, distinct pleural furrow which extends close to abaxial end before being truncated by posterior edge of articulating facet. It divides pleura into a narrower anterior band and a wider posterior band. Posterolateral corner of pleura bluntly angular.

Pygidium subparabolic, about twice as wide (trans.) as long (sag.) Axis about $\frac{1}{4}$ pygidial width anteriorly, rather narrow and composed of eight rings, clearly defined by ring furrows which become progressively shallower towards posterior. Pleural areas with six pairs of rather weakly defined ribs which curve gently backwards and in which pleural furrows are distinctly deeper than the interpleural, the latter running less strongly backwards. Both are truncated a short distance from the margin by the poorly developed pygidial border.

Sculpture of minute granules.

Measurements.

Cranidia	A	A_1	A_2	A_3	A_4	K	δ–δ	
TCD 9619 (E/I)	8·0	5·1	0·8	0·8	1·1	3·7	(4·6)	HOLOTYPE
TCD 9622 (E/I)	—	6·9	—	—	1·2	4·4	—	PARATYPE
TCD 9621 (E/I)	6·2	4·2	0·5	0·5	1·0	3·5	—	

Pygidium	Z	Y	W	X	
TCD 9624 (E/I)	3·9	3·3	(7·9)	2·0	PARATYPE

Remarks. C. strabismus differs from *C. depressus* in the following ways: the preglabellar field is as long (sag.) as the anterior border, the cephalic border is narrower, the pygidial axis has eight rings (compared with five in *depressus*) and the sculpture is of fine rather than coarse granules.

In the collections of the British Museum (Natural History) there is a small number of *Cyphoproetus* specimens collected by Barrande and included under the single number BM 42362. This material, comprising two complete specimens, two cephala and one pygidium is from the Liteň

Formation (late Wenlock) of Loděnice, Prague district, Czechoslovakia. It does not belong to *C. depressus*, but is very similar to the specimens described here as *C. strabismus*, differing in having a more incurved lateral margin of the cephalon, a longer, apparently grooved, genal spine, in having only seven pygidial axial rings and five pairs of pleural ribs and in having a coarse granular sculpture. A new species, closely related to *C. strabismus*, might be represented.

Subfamily SCHIZOPROETINAE Yolkin, 1968
Genus **SCHIZOPROETUS** R. Richter, 1912

Type species. By monotypy; *Proetus celechovicensis* Smyčka, 1895, p. 11; from *Stringocephalus* Limestone (Devonian, Givetian Stage), Čelechovice, Moravia, Czechoslovakia.

Schizoproetus aff. **delicatus** (Hedström, 1923)   Pl. 15, figs. 16, 17

Material, horizon and locality. Five pygidia: GSM 33122, from Wenlock Limestone (Silurian), unspecified horizon, Dudley; NMW 71.6G.247, NMW 72.18G.49–51, from Wenlock Limestone, Nodular Beds, large bedding plane exposure on W side of Wren's Nest Hill, 230–270 yd SW of 'Caves' public house, Dudley, Worcestershire (SO 9350 9210).

Description. Pygidium moderately vaulted with subparabolic outline and with narrow border. Axis anteriorly occupies 33–40% of total pygidial width, and tapers gently backwards to a blunt point. Nine axial rings defined by very shallow ring furrows, the four anterior of which are arched very gently backwards sagittally. At abaxial ends of each ring is a smooth, weakly depressed muscle area. Axial furrow narrow and distinct, shallowing near posterior end of axis, which does not reach pygidial border. Adaxial section of pleural areas nearly horizontal in transverse section, abaxial section steeply declined. Eight pairs of pleural ribs with scalloped profile, which widen very slightly abaxially. Pleural furrows narrow and deep adaxially, shallowing, widening and turning more strongly backwards where pleural areas become steeply declined. Interpleural furrows narrow and very shallow, running approximately parallel with pleural furrows. Anterior and posterior pleural bands of about equal width (trans.). Sculpture of fine granules. Terrace lines on margin.

Measurements.

	Z	Y	W	X
NMW 71.6G.247 (E)	5·8	4·8	(8·8)	(3·2)
NMW 72.18G.49 (E)	6·9	5·6	(10·6)	3·4
GSM 33122 (E)	7·4	6·5	(9·8)	4·2

Remarks. Hedström (1923, p. 4, pl. 1, figs. 1–15) described and figured *Proetus delicatus* from the Halla Beds (late Wenlock) of Hörsne, Gotland. On examination of the type and abundant additional material of this species, I consider it to be referable to *Schizoproetus*, and it is to be redescribed in a forthcoming paper.

The rare pygidia from Dudley closely resemble those of *S. delicatus*, but differ in having nine (rather than 10–11) axial rings, in having differently shaped muscle-areas on the axis and in having a sculpture of fine granules (in contrast to pits on similarly sized specimens of *S. delicatus*). The Dudley specimens probably represent a new species, but the present material is insufficient on which to erect one, and so is referred to as *S.* aff. *delicatus*.

Genus **CRASSIPROETUS** Stumm, 1953

Type species. Original designation; *Proetus* (*Crassiproetus*) *traversensis* Stumm, 1953, p. 110, pl. 1, figs. 1, 2, 10–15, 17; from Middle Devonian, Traverse Group, Michigan, U.S.A.

Remarks. Richter, Richter & Struve (*in* Moore 1959, p. O385) placed *Crassiproetus* with doubt in the Proetinae, while Osmólska (1970, p. 155) proposed the subfamily Crassiproetinae to accommodate it and the Carboniferous genus *Conophillipsia* Roberts, 1963; but an association between these genera seems unlikely. The long (sag.) multisegmented pygidium invites comparison with *Schizoproetus* and *Schizoproetoides* Ormiston, 1967, although the cephalic structure does not closely resemble that of either. Comparatively early (Silurian) species of *Schizoproetus* lack the deep glabellar furrows characteristic of later ones and *Crassiproetus* may originate in the Silurian (see below). Thus it may be possible to derive both *Schizoproetus* and *Crassiproetus* from an ancestor with a smooth glabella and multisegmented pygidium. Consequently, but with reservation, *Crassiproetus* is placed in the Schizoproetinae.

Crassiproetus? curtisi nom. nov.   Pl. 7, fig. 7

Name. In honour of Dr. M. L. K. Curtis, who first described this species.

non 1847   *Proetus asaphoides* Hawle & Corda, p. 78.
  1958   *Proetus asaphoides* Curtis, p. 140, pl. 29, fig. 2. (*non* Hawle & Corda, 1847).
  1970   "*Proetus asaphoides*" Curtis; Přibyl, p. 108 (*non* Hawle & Corda, 1847).
  1972   *Proetus asaphoides* Curtis; Curtis, p. 32 (*non* Hawle & Corda, 1847).

Holotype. GSM GSb4687, Pl. 7, fig. 7; a large, incomplete internal mould; figured Curtis 1958, pl. 29, fig. 2; from base of Tortworth Beds (Llandovery Series, Telychian Stage), Cullimore's Trap Quarry, Charfield, Gloucestershire (ST 7203 9623). The only specimen known.

Diagnosis. Glabella ovate, apparently lacking impressed furrows; occipital ring with poorly defined lateral lobes; pygidium parabolic, markedly longer (sag.) than thorax; pygidial axial rings defined by very shallow ring furrows; pleural areas with distinct, shallow, parallel pleural furrows, pleural ribs scalloped in longitudinal section.

Description. There is nothing to be added to the adequate description of Curtis (1958, p. 140).

Remarks. The ovate glabella, apparently without incised furrows, and the large parabolic pygidium invite comparison with the type species of *Crassiproetus*. As the Llandovery specimen is badly preserved, generic assignment here can only be tentative, but it does indicate that *Crassiproetus*-like trilobites were already in existence quite early in Silurian times.

Přibyl (1970, p. 108) considered that *curtisi* might possibly belong to *Latiproetus* Lu, 1962, but its long (sag.) multisegmented pygidium, ovate glabella and thorax of nine rather than ten segments distinguish it from the type species of that genus, which is *Proetus latilimbatus* Grabau, 1924 (Lu 1962, pl. 1, figs. 3–6).

As the specific name *asaphoides* is preoccupied, the new name *curtisi* is substituted here.

Subfamily CORNUPROETINAE R. & E. Richter, 1956
Genus **CORNUPROETUS** R. & E. Richter, 1919

Type species. Originally designated by R. & E. Richter 1919, p. 46; *Proetus cornutus* Goldfuss, 1843, p. 558, pl. 5, fig. 1; from Middle Devonian, Eifelian Stage, Gerolstein, Eifel district, western Germany.

Cornuproetus peraticus sp. nov.   Pl. 7, figs. 3–6, Text-fig. 6

Name. From Greek 'peratikos', meaning alien or foreign. *Cornuproetus* is rare in Britain, but common in other areas, e.g. Czechoslovakia.

Holotype. GSM Zs572, cranidium; Pl. 7, figs. 3a–d; from Dolyhir Limestone (Wenlock Series), Old Radnor district (exact locality unknown), Radnorshire.

Material, horizon and locality. NMW 72.24G.1 free cheek, NMW 72.24G.2–3, pygidia, all from shale band near base of Dolyhir Limestone, exposure beside track of disused railway, 560 yd at 280° from Dolyhir Bridge and 1400 yd SW of church at Old Radnor, Radnorshire (SO 2405 5823). No further material of this species is known.

Diagnosis. Cephalic border broad, convex; preglabellar field $\frac{1}{3}$ length (sag.) of anterior border; pygidial axis with 5 rings, pleural areas with 4 pairs of ribs. Sculpture of rather coarse granules, covering entire cephalon apart from border, and covering entire pygidium.

Description. Cranidium with sagittal length 1·65 times transverse width across δ–δ. Glabella approximately as long (sag.) as wide (trans.), rather weakly inflated, being gently convex in sagittal and transverse profiles. It tapers gently forwards, with bluntly rounded frontal lobe and weakly constricted a short distance behind γ. All lateral glabellar furrows non-impressed smooth areas interrupting granular sculpture. 1p: lozenge shaped, abaxial end opposite anterior section of palpebral lobe, directed obliquely backwards at about 40° to an exsagittal line, dying out well before reaching occipital furrow. Small auxiliary impression associated. 2p: less conspicuous than 1p, abaxial end opposite γ; furrow clavate, expanding adaxially and directed backwards at about 70° to an exsagittal line. 3p: small ovate area, isolated from axial furrow, and situated not far in front of 2p.

Occipital furrow deep, anterior slope nearly vertical, posterior slope much less steep. Median section arched very weakly forwards, lateral ends turn more strongly forwards. Occipital ring a little wider sagittally than laterally, is a little longer (sag.) than preglabellar area and marginally wider (trans.) than glabella. No lateral lobes, but a small median tubercle is developed, placed slightly towards anterior.

Preglabellar field about $\frac{1}{3}$ length (sag.) of convex anterior border, from which it is separated by a broad, shallow anterior border furrow. Section β–γ of anterior branch of facial suture diverges abaxially forwards at 7° from γ, which is a wide angle (c. 145°), close to axial furrow. Palpebral lobe crescentic, inclined at about 20° from axial furrow in transverse section, and a little under $\frac{1}{2}$ sagittal length of glabella. Posterior branches of facial sutures with ε and ξ as independent angles, the short, intervening stretch close to and parallel with axial furrow.

Free cheek with moderately broad, convex field. Lateral border and lateral border furrow like anterior. Posterior border furrow narrower and deeper than lateral, truncated abruptly at base of genal spine. Median groove of rather narrow-based genal spine very short. Posterior border a little narrower than lateral.

Thorax and hypostome unknown.

Pygidium about twice as wide (trans.) as long (sag.), without border. Axis tapers gently backwards and composed of five rings plus a short end-piece, and these are defined by shallow but distinct ring furrows which are arched gently forwards. Axis about 82% of pygidial length and 33% its anterior width, and in longitudinal profile is strongly convex. No postaxial ridge. Pleural areas broad, weakly convex with four pairs of ribs of imbricate profile. Pleural and interpleural furrows of about equal depth, both reaching close to margin, the pleural furrows running more strongly backwards.

Sculpture of fine granules, covering entire cranidium except areas of lateral glabellar furrows and marginal areas of cephalic border. They are finer and less dense on posterior part of anterior border and on preglabellar field (see Pl. 7, fig. 2). Anterior part of anterior border with fine parallel terrace lines.

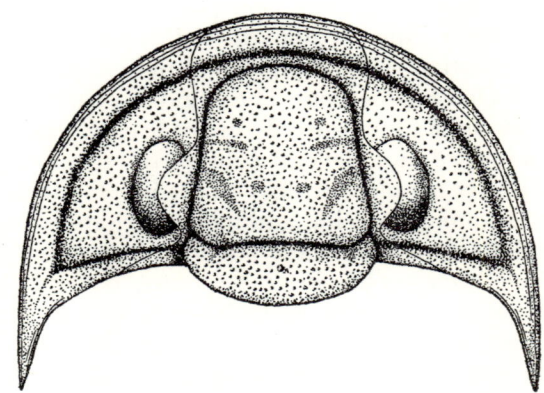

TEXT-FIG. 6. Reconstruction of the cephalon of *Cornuproetus peraticus* sp. nov. (based on Pl. 7, figs. 3, 4). The genal spine shown on this reconstruction is not seen clearly on the plate as it was mostly lost during preparation. ×6 approx.

Measurements.

Cranidium	A	A$_1$	A$_2$	A$_3$	A$_4$	K	δ–δ
GSM Zs572 (E)	6·6	3·9	0·3	0·9	1·5	4·0	(4·8)

Pygidia	Z	Y	W	X
NMW 72.24G.2	5·3	4·3	(10·0)	3·3
NMW 72.24G.3	3·5	2·9	7·4	2·2

Remarks. Five species of *Cornuproetus*, *C. vertumnus* Prantl & Vaněk, 1958 (in Horný, Prantl & Vaněk, 1958), *C. reussi* (Hawle & Corda, 1847), *C. venustus* (Barrande, 1846), *C. consobrinus* Přibyl, 1965 and *C. intermedius* (Barrande, 1846) have been described from the late Wenlock and early Ludlow of the Prague district, Czechoslovakia and one, *C. walliseri* Alberti, 1967, from the early Ludlow of NW Morocco. All these species differ from *C. peraticus* in the type of sculpture, and where pygidia are known, all have four axial rings and three pairs of pleural ribs, compared with five and four respectively in *C. peraticus*.

C. peraticus is the earliest known species of *Cornuproetus* and is the sole representative of the genus in the British Silurian. By contrast, the genus is well represented in the Silurian of the Prague district, Czechoslovakia, but is unknown in Scandinavia.

Subfamily TROPIDOCORYPHINAE Přibyl, 1946
(=Prionopeltiinae Přibyl, 1946; Proetidellinae Hupé, 1953;
Decoroproetinae Erben, 1966)

Genera included. *Tropidocoryphe* Novák, 1890; *Alberticoryphe* Erben, 1966; *Astroproetus* Begg, 1939; *Astycoryphe* Richter & Richter, 1919; *Decoroproetus* Přibyl, 1946; *Denemarkia* Přibyl, 1946; *Hollardia* Alberti, 1964; *Latiproetus* Lu, 1962; *Paraproetus* Přibyl, 1964; *Phaseolops* Whittington, 1963; *Pribylia* Erben, 1951; *Prionopeltis* Hawle & Corda, 1847; *Stenoblepharum* Owens, 1973; *Wolayella* Erben, 1966, and doubtfully *Phyllaspis* Richter, 1863 and *Proetina* Přibyl, 1946.

Diagnosis. Glabellar furrows absent, or only weakly incised; preglabellar field always present; tropidium commonly present; lateral occipital lobes rarely well developed; rostral plate triangular or trapezoidal, connective sutures converge backwards; thorax of 9 or 10 segments, without preannulus; pygidial axis with 3–10 rings, postaxial ridge may or may not be developed; pleural areas with 3–9 pairs of ribs which are of imbricate profile (Text-fig. 2); sculpture characteristically of striations, but may be granular or smooth.

Remarks. Richter, Richter & Struve (*in* Moore 1959, p. O395 and p. O397) recognized the Proetidellinae (range Ordovician and Silurian) and the Tropidocoryphinae (range Silurian and Devonian) as two distinct subfamilies. The type genus of the former is *Proetidella*, which has been claimed by Owens (1973) to be a junior synonym of *Decoroproetus*, which Richter, Richter & Struve (*op. cit.*) assigned with question to the Tropidocoryphinae, but which Erben (1966, p. 170) placed in a new subfamily, the Decoroproetinae. The pygidial pleural rib-structure of *Decoroproetus* is very similar to that of *Tropidocoryphe*, and the only major feature distinguishing the two genera is the presence of the tropidium in the latter. Several *Decoroproetus* species (e.g. *D. scrobiculatus* sp. nov. —see Pl. 9, figs. 12, 15, 16) have tropidial ridges (Text-fig. 1); these are commoner in later (Upper Silurian and Lower Devonian) species, and might perhaps be regarded as being incipient tropidia. The Tropidocoryphinae *sensu* Richter, Richter & Struve (*in* Moore, p. O397) and Erben (1966) apparently have their origins in *Decoroproetus*, but it is by no means certain that all these genera arose from one stock leading from it—indeed, it is far more likely that several independent lines were involved, arising by iterative evolution. Because of the morphological similarities between *Decoroproetus* and other tropidocoryphines and their probable phylogenetic relationship, a separation at subfamily level seems quite artificial, and hence the Proetidellinae and Decoroproetinae are considered to be synonymous with the Tropidocoryphinae.

Erben (1966) has published the first part of a revision of the Tropidocoryphinae, and differentiated its members from those of the Proetidellinae on, among other characters, the type of pygidial pleural ribs. He considered proetidelline pleural ribs to be exemplified by those of *Prantlia* (see Erben 1966, p. 180, and text-fig. 1), of which he had specimens at his disposal (Professor Dr. H. K. Erben, personal communication, 1969), and was misled by the inaccurate illustration of *Proetidella* in the *Treatise* (p. O396, fig. 301, 1), whose pygidial pleural ribs he presumed to be like *Prantlia*. The pygidial pleural rib-structure of *Prantlia* is like that of *Warburgella* Reed, 1931, but unlike that of *Proetidella* (=*Decoroproetus*). Besides their pleural rib-structure *Prantlia* and *Warburgella* also differ from *Decoroproetus* and other tropidocoryphines in another very important respect —the connective sutures of the rostral plate *diverge* (see Text-fig. 1B) instead of *converge* backwards. On the basis of these distinctive characters, I include *Warburgella* and *Prantlia* in a new subfamily, the Warburgellinae.

Genus **DECOROPROETUS** Přibyl, 1946

(=*Warburgaspis* Přibyl, 1946; *Proetidella* Bancroft, 1949; *Ogmocnemis* Kielan, 1960)

Type species. Originally designated by Přibyl 1946, p. 92; *Proetus decorus* Barrande, 1846, p. 64; from Liten Formation (Wenlock, *flexilis* Zone), Loděnice, Prague district, Czechoslovakia.

Diagnosis. Lateral glabellar furrows commonly weak or absent; preglabellar field typically sigmoidal in longitudinal section, but may be concave or straight; tropidium absent; lateral occipital lobes commonly absent, or weak in some species; eye socle commonly distinct, with lower margin in some cases defined by incised furrow; thorax of 10 segments; pygidial axis with 5–10 rings, pleural areas with 4–6 pairs of ribs whose pleural furrows deepen and curve more strongly backwards abaxially; postaxial ridge commonly present; sculpture of continuous or discontinuous striations, covering entire exoskeleton, or localized.

Remarks. This genus is discussed fully by Owens (1973, p. 134).

Species	Eye socle	Profile of preglabellar field		Number of pygidial axial rings	Sculpture
		sigmoidal	concave		
fearnsidesi		x		5–6	continuous striations
calvus		x		?5	"
jamesoni		x		4	"
piriceps	or	x		5–7	"
papyraceus		x		6–8	"
aff. subornatus		x		5	continuous striations and granules
asellus			x	6	continuous striations
cf. evexus			x	6–7	?
scrobiculatus		x		7–9	continuous striations

TABLE 4. Summary of diagnostic characters of *Decoroproetus* species described herein.

1. **Decoroproetus fearnsidesi fearnsidesi** (Bancroft, 1949) Pl. 7, figs. 8–12

1878	*Proetus? ovatus* MS; Huxley, Newton & Etheridge, p. 42 [*nom. nud.*].
1949	*Proetidella fearnsidesi* Bancroft, p. 304, pl. 10, fig. 23.
1953	*Decoroproetus fearnsidesi* (Bancroft); Přibyl, p. 60.
1958	*Decoroproetus fearnsidesi* (Bancroft); Dean, pp. 201, 219.
1959	*Proetidella fearnsidesi* Bancroft; Moore, fig. 301, 1, p. O396.
1962	*Proetidella fearnsidesi* Bancroft; Dean, p. 126.
1963	*Proetidella fearnsidesi* Bancroft; Dean, p. 243, pl. 45, figs. 3–8, 12, 14.
1963a	*Proetidella fearnsidesi* Bancroft; Dean, p. 55.
1964	*Proetidella fearnsidesi* Bancroft; Dean, p. 275.
1966a	*Astroproetus fearnsidesi* (Bancroft); Whittington, p. 81.
1969	*Proetidella fearnsidesi* Bancroft; Pillet, p. 73, pl. 4, fig. 8, pl. 6, fig. 25.
1970	*Decoroproetus fearnsidesi* (Bancroft); Owens, p. 315.

Holotype. By monotypy; BM In42083, Pl. 7, fig. 8; a complete external mould, figured Bancroft 1949, pl. 10, fig. 23, refigured Dean 1963, pl. 45, fig. 3; from Smeathen Wood Beds (Caradoc Series, Harnagian Stage, *Reuscholithus reuschi* Zone), exposure in old cartway 70 yd N of extreme S end of Smeathen Wood, Horderley, Shropshire (SO 4068 8539).

Material, horizons and localities. Numerous specimens, some complete, from the type locality include BM In50662, In51449–54, In51522–23, In55449, GSM JD59, 64, 80, 86, 92. A few complete specimens and detached exoskeletal parts from same horizon as holotype, including BM In50616, BM In51165, BM In51455, NMW 71.6G.25–28, exposure in field 550 yd NW of Woolston House, Woolston, Shropshire (SO 4215 8745), GSM DEW342 from outcrop 340 yd at 350° from the Hough, Little Stretton (SO 4505 9148), GSM Mi426 from hill track 1800 yd at 20° from Cwms Farm, Caer Caradoc (SO 4804 9550). An almost complete internal mould, BM In51152, from Costonian Stage, *Harknessella subquadrata* Zone, old quarry just W of W end of Round Nursery, S of Harnage Grange, Shropshire. Incomplete cranidium, NMW 72.42G.1, and pygidium, NMW 72.42G.2, from basal(?) Caradoc, quarry 100 yd WSW of Lampeter House, Lampeter Velfrey, Pembrokeshire (SN 1507 1438). Pygidium NMW 72.42G.3, and free cheek, NMW 72.42G.4, from high Llandeilo or low Caradoc, old quarry 800 yd NW of Llan-mill, Pembrokeshire (SN 1341 1438).

There are also several poorly localized specimens (e.g. BM I3199, GSM 35613–17 ('syntypes' of *Proetus? ovatus*) 35619–20) whose matrix is of a similar lithology to the Smeathen Wood Beds, and which probably originate from the Horderley district (see Dean 1963, p. 245).

Diagnosis. Glabellar outline may be very weakly constricted, frontal margin well rounded; cephalic border furrow weak, border ill-defined; lateral glabellar furrows inconspicuous, not incised; preglabellar field sigmoidal in longitudinal profile; lower margin of eye socle diverges from upper at either end, but not defined by incised furrow; occipital ring without lateral lobes, pygidial axis with 6 rings, small postaxial ridge present; pleural areas with 5 pairs of ribs; sculpture of continuous striations, covering entire exoskeleton.

Description. This species has been adequately described by Dean (1963, p. 243).

Measurements.

Cranidia	A	A_1	A_2+A_3	A_4	K	δ–δ	
BM In42083 (E)	6·0	3·7	1·3	1·0	4·7	5·8	HOLOTYPE
BM I3199 (I)	7·9	5·1	1·6	1·2	4·9	5·3	
BM In51522 (I)	7·0	4·5	1·3	1·2	4·5	—	
BM In55449 (I)	6·1	3·9	1·2	1·0	4·2	—	
BM In51455 (I)	5·2	3·1	1·3	0·8	2·9	—	

Pygidia	Z	Y	W	X	
BM In42083 (E)	4·1	3·7	9·6	2·4	HOLOTYPE
BM I3199 (I)	6·1	5·5	9·3	3·0	
NMW 72.42G.3 (E)	—	—	7·3	2·4	
NMW 72.42G.2 (E)	4·4	3·3	7·2	2·2	
BM In51454 (I)	1·4	1·1	3·3	1·0	

Remarks. Comparative remarks are made under *D. calvus* (see below), which is evidently closely related to *D. fearnsidesi*. Dean (1963, p. 245) identified *D. fearnsidesi* from beds belonging to the late Costonian and early Harnagian Stages. All the post-Harnagian specimens compared with *fearnsidesi* by Dean are here considered to belong to *D. calvus*.

The free cheek from Llan-mill, NMW 72.42G.4 (Pl. 7, fig. 10) has an eye socle whose lower margin diverges strongly from the upper at either end, and has its median section bowed adaxially. One specimen from Woolston (BM In51165) has a similar eye socle, but in most specimens of *fearnsidesi* the eye socle is not so well developed. Its poor development in many specimens may be due to imperfect preservation, or possibly to original variation within the population.

2. **Decoroproetus fearnsidesi** (Bancroft, 1949) **pristinus** subsp. nov. Pl. 7, figs. 13–19

Name. From the Latin 'pristinus', meaning early; this is the earliest known British *Decoroproetus*.

1963 *Proetidella* sp.; MacGregor, p. 795, pl. 116, figs. 11–13.

Holotype. NMW 72.4G.1a, b, complete internal mould with counterpart external mould; Pl. 7, figs. 15a, b; from Middle? Llandeilo Beds (Llandeilo Series), stream section in Afon Cîb, 150 yd SW of ford and 700 yd WNW of Maes-y-fallen, approximately 1 mile SE of Llandeilo, Carmarthenshire (SN 6448 2120).

Material, horizon and localities. SM A46912, cranidium, from 200 yd N of Nant (SJ 1217 2815), and NMW 63.45G.57, cranidium, SM A46911, free cheek, SM A53010, pygidium from 80 yd N of Nant (SJ 1227 2807), all from Upper Llandeilo, *c.* 1 mile N of Llanrhaiadr-ym-Mochnant, Montgomeryshire. Numerous silicified specimens, e.g. BM In56706–7, BM In56710–15, BM In56730–34, from Lower Llandeilo, old quarry 300 yd SSE of Llangwm Farm, 1½ miles NW of Llandeilo (SN 6083 2385). From Llandeilo Series, exact horizon uncertain: BM It8865, small cranidium from old quarry 340 yd S of Cwm-agol farm, 1½ miles SW of church at Llangathen, Carmarthenshire (SN 5645 2071); NMW 72.42G.5–7, cranidia, NMW 72.42G.8–10, free cheeks and NMW 72.42G.11–13, pygidia from outcrops S of Dryslwyn Castle, *c.* 2¼ miles SW of church at Llangathen, Carmarthenshire (SN 5543 2023); several small silicified specimens including BM It8862–64 from Dynevor Park, Llandeilo, exact locality unspecified. OUM B256–7, complete specimens, have no horizon or locality information, but are almost certainly from the Llandeilo Series, probably from the Llandeilo district.

Diagnosis. As for *D. fearnsidesi fearnsidesi* except: border fairly well defined, pygidial axis with 5 rings, pleural areas with 4 pairs of ribs.

Measurements.

Cranidia	A	A₁	A₂+A₃	A₄	K	δ–δ	
NMW 72.4G.1a (I)	5·7	4·0	1·0	0·7	3·7	(4·5)	HOLOTYPE
NMW 63.459G.57 (I)	3·9	2·3	0·8	0·8	2·1	3·0	
OUM B257 (I)	3·5	2·3	0·7	0·5	2·5	3·2	
OUM B256a (I)	3·1	2·0	0·7	0·4	2·0	(2·3)	

Pygidia	Z	Y	W	X	
NMW 72.4G.1a (I)	3·8	2·9	6·9	2·1	HOLOTYPE
OUM B257 (I)	1·9	1·4	5·2	1·3	
NMW 72.42G.11 (E)	(2·4)	(1·7)	3·6	1·3	
OUM B256a (I)	1·5	1·2	3·7	1·0	

Remarks. At several localities in the Llandeilo district limestones of Llandeilo age have yielded a rich fauna of silicified trilobites, among them a number of specimens of *D. fearnsidesi pristinus*, and included in these are a number of growth stages (e.g. Pl. 7, figs. 13, 16, 17). These specimens, together with four of *D. scrobiculatus* (Pl. 10, figs. 2, 4, 6, 7) are some of the few early ontogenetic stages of *Decoroproetus* known. It is of interest to compare them with similar sized specimens of *Proetus pluteus* (Whittington & Campbell, 1967, pl. 1, figs. 23–4, 30–1; pl. 3, figs. 6–8, 14–16) from the Silurian of Maine, and *Proetus talenti* (Chatterton, 1971, pl. 14, figs. 1–3) from the Devonian of New South Wales. As yet, ontogenetic stages are but poorly known in proetids, with the most complete suites being those mentioned above.

3. Decoroproetus calvus (Whittard, 1961)   Pl. 8, figs. 1–8

1911	*Proetus* aff. *girvanensis* Nicholson & Etheridge; Wade, p. 429.
1958	*Decoroproetus?* sp.; Dean, p. 220.
1959	*Proetidella* aff. *fearnsidesi* Bancroft; Dean, p. 206.
1961	*Ogmocnemis calvus* Whittard, p. 186, pl. 24, fig. 15.
1962	*Proetidella?* *marri* Dean, p. 124, pl. 16, figs. 4, 6, 9; pl. 17, figs. 5, 6, 8, 9.
1963	*Proetidella* cf. *fearnsidesi* Bancroft; Dean, p. 245, pl. 45, figs. 9–11.
?1963a	*Proetidella?* sp.; Dean, p. 55, pl. 1, figs. 8, 9.
1966a	*Astroproetus* cf. *fearnsidesi* (Bancroft); Whittington, p. 82, pl. 25, fig. 12; pl. 26, figs. 2, 3, 5.
1966a	*Astroproetus marri* (Dean); Whittington, p. 83.
1969	*Astroproetus* aff. *fearnsidesi* (Bancroft); Romano & Diggens, p. 605.
1970	*Astroproetus? marri* (Dean, 1962); Ingham, p. 30.

Holotype. GSM 87169, Pl. 8, figs. 1a, b; internal mould of cranidium with counterpart external mould, figured Whittard 1961, pl. 24, fig. 15; from Ordovician, Caradoc Series, Soudleyan Stage, *Broeggerolithus broeggeri* Zone, near base of Whittery Shales, Whittery Quarry, at S end of Whittery Wood, 0·9 mile ESE of church at Chirbury, Shropshire (SO 2746 9808).

Material, horizons and localities. This species has been recorded from the post-Harnagian part of the Caradoc Series. From the Soudleyan Stage: cranidium BM In34326 and free cheek, BM In34325 from *B. broeggeri* Zone?, roadside quarry 150 yd NE of Chatwall Farm, Chatwall, Shropshire (SO 5137 9743); cranidia BM In51503–4, from *B. broeggeri* Zone, exposure behind Glenburrel farmhouse, Horderley, Shropshire (SO 4132 8621); cranidium BM In51456 from *Broeggerolithus soudleyensis* Zone, roadside quarry 100 yd SE of Glenburrell Farm, Horderley, (SO 4143 8611); ill-preserved pygidium BM In54842, probably of *D. calvus*, from Soudley Sandstone, old quarry 50 yd SE of Soudley Pool, Soudley, Shropshire (SO 4771 9183); cranidia, free cheeks and pygidium, GSM FGD1136, 1138, 1140, 1144 from Chatwall Flags, River Onny, 2130 yd at 284° from Wistanstow church (SO 4132 8610); cranidia and free cheeks, GSM Mi394–7, 401, 403–4 from Chatwall Flags, road section at Willstone, 960 yd at 107° from Caer Caradoc summit, 42 ft above base of section (SO 4858 9512); cranidium, ICMM 5359 from Gaerfawr Grits, quarry at W end of Moel-y-Garth Hill, 2–2½ miles NW of Welshpool, Montgomeryshire. Probably Soudleyan Stage; cephalon with parts of five thoracic segments, NMW 62.24G.42.1 from Meifod (exact locality unspecified), Montgomeryshire; cranidium, BM In57979, 1195 yd at 354° from Lion Hotel, Meifod. From Lower Longvillian Substage: cranidia BM In54644 (holotype of *P.? marri* Dean), BM In54645–47, GSM PJ3617, free cheeks, BM In54648, BM In55881, GSM PJ3616, pygidia BM In55880, In55882, all from *corona* Beds, Dufton Shales, *Bancroftina typa* Zone, exposure in Harthwaite Sike, 1450 yd E of Billysbeck Bridge, Dufton, Westmorland (NY 7070 2473) [Dean 1962, p. 75, fig. 4, loc. E3]. From Longvillian (undifferentiated): cranidium, from Glanrafon Beds, roadside exposure 2⅛ miles SW of Dolwyddelan, Caernarvonshire (SH 7080 5029); cranidium, BU 926 and cranidium BU 927 (figd. Whittington 1966a, pl. 25, fig. 12; pl. 26, figs. 2, 3, 5), calcareous ashes of Gelli-grîn Group, below Cymerig Limestone, 100 ft above base of section on Ffrîdd Bâch, S of Maes-meillion, 3½ miles S of Bala, Merioneth (SH 925 305).

A cranidium, BM In56799 and a free cheek, BM In56787, possibly belonging to this species have been figured by Dean (1963a, pl. 1, figs. 8, 9) from Actonian Stage, Stile End Beds, from his localities λG and λE, 2300 ft at 348° and 2000 ft at 325° from Stockdale Bridge, Long Sleddale, Westmorland.

Diagnosis. Glabella weakly constricted, frontal margin bluntly rounded or transverse; weak lateral glabellar furrows seen with good preservation; occipital ring without lobes; preglabellar field sigmoidal in profile, anterior border well-defined, upturned; eye socle with non-incised lower margin which diverges from upper at either end; sculpture of fine, continuous striations.

Description. Cephalon with narrow, upturned border defined by shallow but distinct anterior and lateral border furrows. Cranidium with sagittal length a little greater than palpebral width. Glabella approximately as wide (trans.) as long (sag.), rather weakly inflated, tapering gently forwards, weakly constricted laterally and with frontal margin bluntly rounded to transverse. Lateral glabellar furrows, when seen (e.g. Pl. 8, figs. 2, 7, 8), are weak. Whittard (1961, p. 186) noted their absence in the holotype, the only specimen known to him, which is badly preserved; Dean (1962, p. 124) stated that *marri* lacked them also, but Ingham (1970, p. 30) suspected their presence on the holotype of *marri* (Pl. 8, fig. 8), and his suspicion is here confirmed. 1p: abaxial end opposite anterior end of palpebral lobe, and from here is directed backwards at about 45°. 2p: close to point of greatest constriction of glabella, opposite γ, directed backwards weakly, shorter than 1p. 3p: close to anterolateral corner of glabella, very weak, seen on Pl. 8, fig. 8.

Occipital furrow broader and deeper than axial and preglabellar furrows, arched weakly forwards sagittally and again laterally, anterior slope almost vertical, posterior slope inclined at about 35°. Occipital ring about as wide (sag.) as preglabellar field, as wide (trans.) as glabella and maintaining constant width laterally. Lateral occipital lobes apparently not developed. Small median occipital tubercle present. Preglabellar field sigmoidal in longitudinal section (Pl. 8, fig. 1b), between $\frac{1}{4}$ and $\frac{1}{5}$ sagittal length of glabella. Section β–γ of anterior branch of facial suture diverges abaxially forwards at 15°–34° from γ, which is close to axial furrow. Posterior branches with ϵ and ξ apparently as a single angle. Palpebral lobe narrow, subcrescentic, between $\frac{1}{3}$ and $\frac{2}{5}$ sagittal length of glabella. Eye crescentic, eye socle not seen well on any of available specimens, but appears to be poorly developed. Field of free cheek broad, weakly convex. Posterior border furrow deep and broad, truncated at base of rather short genal spine.

Most complete thorax has parts of five segments preserved, which are similar to those of *D. fearnsidesi* (cf. Pl. 8, fig. 2, and Pl. 7, fig. 8).

The most complete pygidium known (Pl. 8, fig. 5) is semi-elliptical in outline. Axis strongly convex in sagittal profile, with probably five axial rings, defined by shallow ring furrows, and a short end piece. Four pairs of pleural furrows seen on most complete specimen, on which only first interpleural furrow is seen and four pairs of moderately strong pleural furrows. Anterior pleural band narrower (exsag.) than posterior.

Sculpture of fine striations, seen on the glabellas of GSM PJ3617 (Pl. 8, fig. 6), and BM In54645 (Dean 1962, pl. 17, fig. 5) and BM In54646 (Dean 1962, pl. 16, fig. 9).

Measurements.

Cranidia	A	A_1	A_2+A_3	A_4	K	δ–δ	
GSM 87169 (I)	9·0	5·5	2·2	1·3	5·8	—	HOLOTYPE
BM In54644 (I)	7·2	4·7	1·5	1·0	4·7	6·2	Holotype of *marri*
GSM PJ3617 (E/I)	5·8	3·4	1·4	1·0	3·3	—	
ICMM 5359 (I)	5·6	3·7	1·3	0·6	3·5	—	
NMW 62.24G.42.1 (I)	5·4	4·0	0·9	0·5	3·5	(4·0)	

Remarks. Whittard (1961, p. 186) distinguished *D. calvus* from *D. fearnsidesi* by its better developed preglabellar field and lack of 1p glabellar furrows. The last feature of *fearnsidesi* was later pointed out by Dean (1962, p. 126) to be the result of crushing in the holotype. Dean (1962, p. 126; 1963, p. 245) has also suggested that *calvus* and *fearnsidesi* might be conspecific, and drew attention to similarities exhibited by cranidia from the Soudleyan of South Shropshire (Dean, 1963, pl. 45, figs. 9–11) and the holotype of *calvus*. *Proetidella? marri* from the Lower Longvillian of the Cross Fell Inlier was described by Dean (1962, p. 124, pl. 16, figs. 4, 6, 9; pl. 17, figs. 5, 6, 8, 9), who distinguished it from *fearnsidesi* by its narrower frontal glabellar lobe, its more transverse frontal margin and its better differentiated preglabellar field and anterior border. There is some range in the glabellar shape in *marri* (cf. Dean 1962, pl. 17, figs. 5 and 8) and some specimens are very similar to the type of *calvus*. Both *calvus* and *marri* can be distinguished from *fearnsidesi* by the well differentiated preglabellar field and anterior border, but it is difficult to distinguish *marri* from *calvus*. On the present evidence, therefore, I prefer to consider *marri* as a synonym of *calvus*.

D. calvus appears to replace *D. fearnsidesi* in the Anglo-Welsh area above the Harnagian Stage, with the youngest known definite examples from the Lower Longvillian, and a possible example from the Actonian Stile End Beds. Another species very similar to *D. calvus* is *D. piriceps* (Ingham, 1970) (see below). For differences from *D. jamesoni*, see 'remarks' under that species and Table 4.

4. Decoroproetus jamesoni (Reed, 1914)   Pl. 8, figs. 9–13, ?14–16

1914	*Cyphaspis jamesoni* Reed, p. 27, pl. 4, fig. 8.	
1925	?'*Phaetonides*' *jamesoni* (Reed); Warburg, p. 186.	
1931	*Cyphaspis jamesoni* Reed; Reed, p. 15.	
1940	*Proetus vicinus* Reed, p. 159, pl. 8, fig. 1.	
1947	*Proetus* (*Eremiproetus?*) *reedi* nom. nov. Přibyl, p. 1, text-fig. 1 (pro *Proetus vicinus* Reed, *non* Barrande, 1872).	
1947	*Proetus ardmillanensis* Begg, p. 42, pl. 3, fig. 3.	
1951	*Proetus balclatchiensis* Begg, p. 362, pl. 1, fig. 1.	
1951	*Proetus trefoileum* Begg, p. 364, pl. 1, fig. 2.	
1951	*Proetus* sp.; Begg, p. 365, pl. 1, fig. 3.	
1952	*Proetus ardmillanensis* Begg; Harper, pp. 89, 107, pl. 5, fig. 3.	
?1962	*Proetidella* sp.; Tripp, p. 13, pl. 2, fig. 15.	
?1967	*Proetidella* sp. A; Tripp, p. 52, pl. 2, figs. 13,14.	
?1967	*Proetidella* sp. B; Tripp, p. 53, pl. 2, figs. 15–17.	
1970	*Decoroproetus jamesoni* (Reed); Owens, p. 315.	

Holotype. By monotypy; BM In21971, Pl. 8, figs. 9a–c; a complete internal mould, with counterpart external mould, figured Reed 1914, pl. 14, fig. 8; from Balclatchie Group (Caradoc Series), Balclatchie, near Girvan, Ayrshire.

Material, horizons and localities. Nearly complete specimen, BM In40485, counterpart internal and external moulds; cephalon with parts of three attached thoracic segments, HM A4122 (holotype of *Proetus balclatchiensis*), external mould of cephalon, HM A4123 (holotype of *P. trefoileum*), cranidia BM In37547 (holotype of *P. vicinus*), HM A4124a,b, BM In37507, BM In40529, all from type locality. Cranidia, HM A3692 (holotype of *P. ardmillanensis*), HM A3825, HM A5056, BM In22648–49, BM In47420 from Balclatchie Group of Dow Hill, near Girvan. Cranidia, NMI 51/12, LU 7003–5, from Caradoc Series, Upper Tuffs and Shales of Grangegeeth Volcanic Series, 1⅜ miles ESE of Grangegeeth, Co. Meath, Ireland (Harper 1952, pl. 7, loc. 26).

Diagnosis. Glabella weakly constricted, subquadrate: preglabellar field with sigmoidal profile; cephalic border distinct, defined by shallow border furrow; lower margin of eye socle not incised, diverging strongly from the upper margin at either end and with the median part more or less exsagittal; pygidium rather short (sag.), axis with 4 rings, pleural areas with ?3 pairs of ribs; sculpture of fine, continuous striations, covering entire exoskeleton.

Description. Cephalon subparabolic, border rather narrow, upturned and defined by shallow but distinct anterior and lateral border furrows. Glabella subquadrate, sagittal length and greatest transverse width nearly equal, lateral margin weakly constricted. Between one and three pairs of simple, weakly impressed lateral glabellar furrows can be seen on some specimens (e.g. Pl. 8, fig. 10), but bad preservation commonly obscures them. Glabella rather weakly inflated in transverse and longitudinal sections (see Pl. 8, fig. 9c).

Occipital furrow deep, flexed forwards weakly sagittally, and more strongly at either end, with both anterior and posterior slopes steep (see Pl. 8, fig. 9c), the former being steeper. Occipital ring narrows slightly laterally, has a distinct median tubercle and lacks lateral lobes. In profile (see Pl. 8, fig. 9c) it is flat and gently inclined towards the posterior. Transversely it is as wide as the glabella.

Preglabellar field with sigmoidal profile in longitudinal section, and between ⅓ and ¼ sagittal length of glabella. Section β–γ of anterior branch of facial suture diverges abaxially forwards at 18°–30° from γ, which is close to anterolateral corner of glabella. Posterior branches with ϵ and ξ a single, rather open angle, close to axial furrow. Palpebral lobe apparently subcrescentic, but is ill preserved on all material seen. Eye close to glabella, posteriorly placed and approaching ½ its sagittal length. Eye socle about ¾ length of glabella, lower margin diverging strongly from upper at either end, its median section running nearly exsagittally.

Field of free cheek weakly convex, strongly declined from eye region. Posterior border furrow deep, truncated at base of genal spine which has a shallow median furrow and extends backwards as far as sixth thoracic segment. Cephalic doublure of similar width to border, ventrally convex and with prominent parallel terrace lines. On holotype (Pl. 8, fig. 9b) mould of triangular rostral plate and the hypostome can be seen.

Thorax of ten segments. Strongly arched axis tapers evenly backwards so that tenth ring is a little under $\frac{1}{2}$ width (trans.) of first. Pleural furrow divides pleura into a broad (exsag.) posterior and a narrow (exsag.) anterior pleural band. Posterolateral corner of pleura angular.

Pygidium small, subparabolic and about twice as wide (trans.) as long (sag.). Axis short (sag.), strongly convex in longitudinal profile, with four rings. Preservation not sufficient on any specimen to show if postaxial ridge is present. Pleural areas apparently have three pairs of ribs.

Exoskeleton has a sculpture of fine, continuous striations.

Measurements.

Cranidia	A	A_1	A_2+A_3	A_4	K	$\delta-\delta$	
BM In21971 (I)	2·5	1·5	0·7	0·3	1·6	(2·1)	HOLOTYPE
BM In40529 (I)	2·0	1·1	0·5	0·4	1·1	—	
BM In37547 (I)	—	2·4	—	0·7	2·3	—	Holotype of *vicinus*
HM A3692 (I)	3·0	1·6	0·9	0·5	1·7	—	Holotype of *ardmillanensis*
HM A4122 (I)	—	1·3	(0·5)	0·3	2·1	(1·8)	Holotype of *balclatchiensis*
HM A4123 (E)	2·9	1·7	0·8	0·4	2·1	(2·3)	Holotype of *trefoileum*
HM A4124a (I)	2·6	1·4	0·7	0·5	1·6	—	

Remarks. Since Reed (1914, p. 27) described *jamesoni*, specimens of *Decoroproetus* from the Balclatchie Group have been described as *Proetus vicinus* Reed, 1940, *Proetus ardmillanensis* Begg, 1947, *Proetus balclatchiensis* Begg, 1951 and *Proetus trefoileum* Begg, 1951. Examination of the types of these species, and a comparison of one with another and with the type of *jamesoni* shows that although there is some variation (in part due to preservation), none of the forms has any character which warrant making it a separate species, and hence all are included in *jamesoni*, the senior synonym.

Tripp (1962, p. 13, pl. 2, fig. 15; 1967, pp. 52–53, pl. 2, figs. 13–17) distinguished three *Proetidella* species from the *confinis* Flags and from the Upper Stinchar Limestone of the Stinchar Valley in the Girvan district. These forms (Pl. 8, figs. 14–16) show some variation in the length–breadth proportions of the glabella and in the length of the preglabellar field. All the specimens are small and ill preserved, but as far as can be seen some or all may well be conspecific with *D. jamesoni*, and are here tentatively assigned to this species.

The short (sag.) pygidium, the type of eye socle and glabellar outline distinguish *D. jamesoni* from *D. fearnsidesi* and *D. calvus* (cf. Pl. 7, fig. 8 and Pl. 8, figs. 1–5 with Pl. 8, fig. 9). Closest resemblance is seen in *D. furubergensis* Owens (1970, p. 312, fig. 6) from Stages 4bγ and 4bδ of the Chasmops Series (Caradoc) of the Oslo region, Norway, but the surface sculpture and type of eye socle differ, although the glabellar shape is similar; the pygidium has similar length-breadth proportions, but has more axial rings (6, compared to 4 in *D. jamesoni*).

5. Decoroproetus piriceps (Ingham, 1970)   Pl. 8, figs. 17–26

1878   *Proetus encrinurus* MS; Huxley, Newton & Etheridge, p. 42 [*nom. nud.*].
1948   *Proetus* sp.; King & Williams, p. 210.
?1966   *Proetidella* sp.; Ingham, p. 497.
1966   *Proetidella* sp. nov.; Ingham, pp. 466, 467, 486, 498, 499.
1970   *Astroproetus? piriceps* Ingham, p. 28, pl. 4, figs. 20–26, 28, 30–32, text-fig. 10, p. 30.
?1970   *Astroproetus? piriceps?* Ingham, p. 30, pl. 4, figs. 27, 29.
?1970   *Astroproetus?* sp.; Ingham, p. 31, pl. 4, fig. 33.
1971   *Astroproetus?* sp.; Dean, p. 35.

Holotype. HUD 2.56, internal mould of incomplete cranidium, figured Ingham 1970, pl. 4, fig. 21; from Ordovician, Ashgill Series, Cautleyan Stage, zone 1, Sally Brow, Murthwaite Inlier, Cautley district, NW Yorkshire (Ingham 1970, text-fig. 3, loc. S 71).

Material, horizons and localities. Besides those in the Cautley district listed by Ingham (1970, p. 28), the following occurrences of this species are known: from Coniston Limestone Group, Applethwaite Beds (Cautleyan Stage, zones 2–3)—cranidia, GSM 35609–10, pygidium, GSM 35611 ('syntypes' of *Proetus encrinurus* Huxley, Newton & Etheridge [*nom. nud.*]) from Troutbeck, Westmorland; cranidia GSM RU3642, RU3604, RU4065, RU4077, free cheeks GSM RU3605, RU4024, RU4044, RU4077 and pygidium GSM RU3636 from Appletreeworth Beck, *c.* 3½ miles SW of Coniston, Lancashire (SD 2465 9285). Cranidia BM It8827, It8828, It8833, free cheeks BM It8829, It8831, It8832, pygidia BM It8830, It8834 from Birdshill Limestone (probably Pusgillian Stage) old quarries 200 yd NW of Birdshill farm, 1¾ miles WNW of Llandeilo, Carmarthenshire (SN 601 231). Cranidia BM It8835, NMW 72.9G.1–2, free cheeks BM It8837, NMW 72.9G.4, pygidia BM It8836, NMW 72.9G.3, 5–6 from Crûg Limestone (?Marshbrookian Stage), old quarry at Crûg Farm, ½ mile NNW of church at Llandeilo (SN 6273 2305). A cranidium, BM In54598 and a free cheek, BM In54599 from lowest Cautleyan Dufton Shales of Billy's Beck near Dufton, Westmorland (Dean 1959, p. 197, fig. 3, loc. D. 2) probably belong to this species.

Diagnosis. Glabella pyriform with three pairs of weakly incised furrows; preglabellar field sigmoidal in profile; occipital ring with weakly developed lateral lobes; eye socle well developed, with lower margin diverging from the upper at either end, with the median section more or less exsagittal; pygidial axis strongly arched with 5–7 rings; no postaxial ridge; 3–5 pairs of pleural ribs, first 2 or 3 pairs of pleural furrows distinct, reaching margin, others faint; sculpture of fine, continuous striations.

Description. See Ingham 1970, p. 28.

Measurements.

Cranidia	A	A_1	A_2+A_3	A_4	K	δ–δ	
HUD. 2.56 (I)	5·1	2·9	1·2	1·0	3·3	(4·8)	HOLOTYPE
GSM RU3604 (I)	7·0	4·3	1·7	1·0	5·0	—	
BM It8833 (E/I)	—	4·0	1·2	—	4·0	—	
BM It8827 (I)	—	4·0	—	0·8	4·0	—	
BM It8835 (E/I)	6·4	3·8	1·4	1·2	4·2	(6·0)	
GSM RU4065 (I)	—	3·7	—	1·1	3·6	—	
BM It8828 (E)	4·2	2·7	0·7	0·8	2·5	(3·2)	

Pygidia	Z	Y	W	X
BM It8834 (E/I)	6·1	4·9	12·0	3·9
BM It8836 (E/I)	5·1	4·0	9·0	3·0
NMW 72.9G.6 (E)	(4·7)	(3·7)	8·9	3·1
BM It8830 (E)	3·8	2·9	6·6	2·1
NMW 72.9G.5 (E)	3·0	2·3	4·7	1·8

Remarks. Ingham (1970, p. 28) based this species on material from the Ordovician inliers of the Cautley district, and all his material was somewhat distorted. Further, less distorted specimens, some retaining the external surface, have since been found in the Coniston and Llandeilo areas. Information from these specimens has necessitated slight amendments to Ingham's diagnosis, and they have confirmed the general accuracy of his (1970, p. 30, text-fig. 10) reconstruction, although it does not show the eye-socle.

D. piriceps is closely comparable with *D. calvus* (Whittard) (=*marri* Dean) as Ingham (1970, p. 30) has suggested, and it can be distinguished from the Caradoc species by its possession of weak lateral occipital lobes, by its more strongly constricted glabella and by its type of eye socle (see Table 4). Specimens from the Crûg Limestone are like those from the Birdshill Limestone (cf. Pl. 8, figs. 17–21, 24, 25 and figs. 22, 23, 26), but the former horizon belongs, according to conodont evidence (Bergström 1964, pp. 51–53), to the Marshbrookian Stage of the Caradoc. However, all other occurrences of *D. piriceps* are from beds of Pusgillian, Cautleyan or supposed Cautleyan age, and further studies on the fauna of the Crûg Limestone may reveal that it, too, is Pusgillian or Cautleyan.

Dean (1963, p. 246, pl. 45, fig. 13) described and figured a proetid cranidium from the Marshbrookian near Acton Scott, S Shropshire as *Proetidella*? sp. This specimen shows similarities to both *D. calvus* and *D. piriceps* in its pyriform glabella (with weak 1p furrows possibly developed), a preglabellar field with sigmoidal transverse profile and a rather narrow anterior border. The last feature is less well defined than in these species, but the variability of either species may turn out to be such as to include Dean's cranidium.

Of foreign species, *D. piriceps* closely resembles *D. brevifrons* (Angelin, 1854) from the Jonstorp Formation of Sweden, particularly in its glabellar outline (Owens 1973, p. 141), but the latter species differs in its smaller eye with the lower margin of the eye socle nearly parallel to the upper and not as distinct as in *D. piriceps*.

6. **Decoroproetus papyraceus** (Törnquist, 1884)   Pl. 9, figs. 1, 2

1884	*Proetus papyraceus* Törnquist, p. 48, pl. 2, figs. 4–6.
1918	*Niobe cunctatrix* Sjöberg, p. 458, pl. 7, figs. 1–2.
1960	*Ogmocnemis irregularis* Kielan, p. 70, pl. 3, figs. 6–9, pl. 4, figs. 8, 9, pl. 26, fig. 1, text-fig. 17, p. 71.
1966a	*Astroproetus irregularis* (Kielan); Whittington, p. 81.
?1966a	*Astroproetus berwynensis?* Whittington, p. 84, pl. 26, fig. 1.
?1966	*Decoroproetus* (*Ogmocnemis*) cf. *irregularis* (Kielan); Ingham, pp. 473, 487, 501.
?1970	*Astroproetus?* cf. *irregularis* (Kielan); Ingham, p. 30, pl. 4, figs. 34–38.
1973	*Decoroproetus papyraceus* (Törnquist, 1884); Owens, p. 149, figs. 8C–H, J, L, M.

Type specimens. From Törnquist's three syntypes, a lectotype has been selected by Owens (1973, p. 150); LPI LO597 T; an internal mould of a cranidium, figured Törnquist 1884, pl. 2, fig. 2 and refigured Owens 1973, fig. 8D; from Fjäcka Shale (Ordovician, Harju Series, *Pleurograptus linearis* Zone), Fjäcka, Lake Siljan district, Dalarne, Sweden. Free cheek LPI LO599t and pygidium LPI LO598t are paralectotypes.

Material, horizons and localities. SM A39017 and A39018, external moulds of cephala with some attached thoracic segments, from upper calcareous shales below Wharfe Conglomerate (Ashgill Series, Rawtheyan Stage, zone 6), just below Wharfe Mill Dam, Austwick Beck, near Austwick, W Riding of Yorkshire (SF 779 696). Specimens described and figured by Ingham (1970, p. 30, pl. 4, figs. 34–38) as *Astroproetus?* cf. *irregularis* (Kielan) from Rawtheyan Stage, zone 6, Cautley Mudstones of the Murthwaite and Westerdale inliers, Cautley district (Ingham 1970, locs. S140, B24, B35), might well belong to this species, but the material is too badly preserved for precise identification. The pygidium, BU 928, figured by Whittington (1966a, pl. 26, fig. 1) as *Astroproetus berwynensis?* from Rawtheyan Stage, Rhiwlas Limestone, beside stream 650 ft at 186° from Eglwys Anne Warren, 2¾ miles NNE of Bala, Merioneth, might also belong here. Whittington (op. cit., p. 84) compared it with *Astroproetus irregularis* (Kielan) [=*D. papyraceus*]. Its pygidial pleural rib-structure suggests assignment to *Decoroproetus* rather than to *Proetus* [=*Astroproetus*] *berwynensis*.

Diagnosis. Glabella tapering gently forwards, very weakly constricted laterally; weakly impressed 1p furrows seen on some specimens; preglabellar field sigmoidal in profile; lower margin of eye socle incised, diverging from upper at either end; pygidial axis with 6–8 rings, pleural areas with 5–7 pairs of ribs; sculpture of fine, continuous striations.

Description. See Kielan (1960, p. 70) for a full description of *Ogmocnemis irregularis*, a junior synonym of *D. papyraceus*, and also Owens (1973) for additional remarks.

Remarks. The affinities of this species have been discussed elsewhere by Owens (1973). In the British Isles, *D. papyraceus* and specimens probably belonging to the species have been recorded only from the Rawtheyan Stage of the Ashgill Series, although it is found in earlier, approximately Pusgillian strata in Sweden. For comparison with other species, see Table 4.

7. **Decoroproetus** cf. **subornatus** (Cooper & Kindle, 1936)   Pl. 9, figs. 3–7

	1932	*Proetus* cf. *kullsbergensis* Warburg; King, p. 104.
cf.	1936	*Proetus subornatus* Cooper & Kindle, p. 364, pl. 52, figs. 16, 24.

Material, horizons and localities. Cranidia, SM A9596, A9598, free cheek, SM A9595 and pygidium, SM A9597 from Ordovician (Ashgill, ?Cautleyan Stage) limestone in Neptunean dyke in Ingletonian, railway cutting 2000 ft S of Horton-in-Ribblesdale Station, W Riding of Yorkshire

(SD 805 720). Cranidium, SM A31744 and free cheek, NMW 71.6G.511 from Keisley Limestone (Ashgill, ?Cautleyan Stage), exposures on Keisley Bank, Keisley, Westmorland (NY 714 249).

Description. Cranidium weakly inflated, glabella subquadrate, as long (sag.) as wide (trans.), tapering gently forwards, weakly constricted opposite γ and with a bluntly rounded frontal lobe. Three pairs of non-impressed glabellar furrows; 1p: crescentic, directed obliquely backwards at about 35°, its abaxial end opposite centre of palpebral lobe and its adaxial end close to occipital furrow; 2p: opposite γ, clavate; 3p: a short distance in front, running weakly forwards.

Occipital furrow deep, anterior slope nearly vertical, posterior slope shallow. Median stretch arched weakly forwards, lateral ends curving strongly forwards. Occipital ring maintains about same width (sag. and exsag.) all the way across, and is markedly wider (trans.) than glabella. Preglabellar field with sigmoidal profile in longitudinal section, a little narrower (sag.) than occipital ring. Section β–γ of anterior branch of facial suture diverges abaxially forwards at 10°–15° from γ, which is close to axial furrow. Posterior branches with ϵ and ξ as a single angle. Palpebral lobe large, crescentic. Eye surrounded with distinct eye socle, whose lower margin diverges strongly from the upper at either end, with its median stretch running nearly exsagittally. Lower marginal furrow not deeply incised. Field of free cheek rather narrow, gently convex. Genal spine long, rather narrow, median furrow distinct only anteriorly, where it is offset abaxially from lateral border furrow.

Pygidium subparabolic, about $1\frac{1}{2}$ times as wide (trans.) as long (sag.) Axis tapers gently backwards and has five rings, defined by apparently shallow ring furrows. No postaxial ridge. Pleural areas with four pairs of ribs. Pleural furrows longer and more strongly curved than interpleural, former reaching margin, latter not.

Sculpture of fine raised striations arranged in a Bertillon pattern on the glabella. On the largest specimen rows of small granules are seen along the striations (see Pl. 9, figs. 3a–c).

Measurements.

Cranidia	A	A_1	A_2+A_3	A_4	K	δ–δ
SM A9598 (E)	4·6	2·6	1·2	0·8	2·6	(3·2)
SM A9596 (E)	7·0	4·1	1·8	1·1	4·0	(5·0)
SM A31744 (E)	7·2	4·0	1·8	1·4	4·1	(4·9)

Pygidium	Z	Y	W	X
SM A9597 (E)	2·8	2·3	4·3	1·8

Remarks. Comparison between the specimens from Horton-in-Ribblesdale and Keisley described above and a plaster cast of a topotype cranidium of *Decoroproetus subornatus* (Cooper & Kindle, 1936) from the Ashgill Whitehead Formation of Percé, Quebec, (kindly sent to the author by Professor P. J. Lespérance of Montreal) indicates little difference between them, but until it is possible to compare other parts of the exoskeleton I prefer to refer to the British specimens as *D.* cf. *subornatus*. The occurrence of the same or very similar species in the three formations suggests that their ages might be similar.

The type of eye socle of *D.* cf. *subornatus* is also seen in such species as *D. jamesoni* (Pl. 8, fig. 9), *D. papyraceus* (Owens 1973, figs. 8E, H, L), *D. piriceps* (Pl. 8, figs. 19–22), and *D.* sp. 2 (Owens 1973, fig. 8N), and this common character may reflect a close phylogenetic grouping. Further comparison of these species is shown in Table 4.

8. **Decoroproetus asellus** (Esmark, 1833)   Pl. 9, figs. 8, 9

1833   *Trilobites asellus* Esmark, pl. 7, fig. 5.
1884   *Proetus modestus* Törnquist, p. 46, pl. 2, figs. 1–3.
1925   *Proetus remotus* Warburg, p. 170, pl. 5, fig. 7.
1947   *Proetus mactaggarti* Begg, p. 40, pl. 3, figs. 1, 2.
1973   *Decoroproetus asellus* (Esmark); Owens 1973, p. 135, figs. 4A, B, D–I (with full synonymy).

Holotype. By monotypy; PMO 56442, a nearly complete internal mould, from Trosviken ved Brevik, S Norway figured by Owens 1973, fig. 4A; Tretaspis Series (Ordovician, Harju Series, Stage 4cα).

Material, horizons and localities. HM A3691 (holotype of *Proetus mactaggarti*), a complete internal mould from Upper Drummuck Group (Ashgill Series, Rawtheyan Stage) a few feet above Starfish Bed no. 3, Lady Burn, Girvan district, Ayrshire; BM In21944, BM In40838, BM In40922, BM In46899, BM In46903, all complete or partially complete internal moulds from Upper Drummuck Group, Starfish Bed no. 1, Thraive Glen, Girvan district. For foreign distribution of this species, see Owens (1973, p. 135).

Diagnosis. Glabella not constricted laterally, frontal lobe well rounded, lateral glabellar furrows weak, not always seen; occipital ring characteristically a little wider (trans.) than glabella; cephalic border narrow, upturned; preglabellar field concave in longitudinal section; eye small, eye socle with incised lower marginal furrow which diverges markedly from upper at either end, ϵ and ξ independent angles; pygidial axis strongly arched longitudinally, with ill-defined rings, pleural areas with 5 pairs of ribs, on which interpleural furrows are inconspicuous; sculpture of fine, continuous striations.

Measurements.

Cranidia	A	A_1	A_2+A_3	A_4	K	$\delta-\delta$	
BM In21944 (I)	4·9	3·8	0·8	0·3	3·1	(3·5)	
HM A3691 (I)	(3·5)	2·8	(0·7)	0·6	3·0	(3·5)	Holotype of *mactaggarti*
BM In40922 (I)	3·5	2·5	0·7	0·3	2·2	(2·5)	
BM In46903 (I)	—	2·0	(0·4)	—	2·5	(3·0)	
BM In46899 (I)	2·1	1·4	0·5	0·2	1·6	(2·0)	

Pygidia	Z	Y	W	X
BM In21944 (I)	3·2	2·5	5·8	1·5
HM A3691 (I)	2·9	2·4	5·1	1·5
BM In40922 (I)	2·2	1·9	4·7	1·3
BM In46899 (I)	1·4	1·0	2·7	0·9

Remarks. In the British Isles, *Decoroproetus asellus* has been recorded only from the Girvan district; its occurrence there is approximately contemporaneous with those in Scandinavia. This species could apparently tolerate a number of different environments—it has been found in reef limestone and dark shale in Scandinavia, and at Girvan occurs in a sandy mudstone.

9. **Decoroproetus** cf. **evexus** Owens, 1973.   Not figured

1969   Proetidae, cranidium type 1; Temple, p. 216, pl. 4, figs. 1–4.
1969   Proetidae and/or Otarionidae free cheeks; Temple, p. 217, pl. 4, figs. 15–16.
?1969   Proetidae, hypostome; Temple, p. 217, pl. 4, figs. 9–11.
1969   Proetidae, pygidium type 1; Temple, p. 217, pl. 4, figs. 13, 20.
1969   Proetidae, pygidium type 2; Temple, p. 218, pl. 4, fig. 17.

Material, horizons and localities. Cranidia, BM It5033–35, free cheeks, BM It5050–51, pygidia BM It5043, It5045, It5048, and possibly hypostomes BM It5040–42 from limestone overlying the Keisley Limestone, exposure near foot of Keisley Bank, Keisley, Westmorland (NY 7136 2377). The horizon is topmost Ashgill (Ingham & Wright *in* Williams *et al.* 1972, p. 47) or basal Llandovery (Temple 1969, p. 199).

Remarks. Owens (1973, p. 148) has recently compared some of the proetids described by Temple (1969, pp. 216–220) with *Decoroproetus evexus* Owens (1973, figs. 5L–N, figs. 7A–I, L) from the late Ordovician of Norway and Sweden, and suggested that they might be conspecific. Those of Temple's specimens listed in the above synonymy appear to belong there, but until better material is available are referred to as *D*. cf. *evexus*. The variation exhibited by Temple's pygidia types 1 and 2 can be seen to occur in *D. evexus* (see Owens, 1973, p. 148), thus confirming Temple's (1969, p. 220) suspicion that one variable species is represented.

The other proetids described and figured by Temple (1969) from the same horizon and locality as 'cranidium type 2' and 'pygidium type 3' are not easy to place. Because the specimens listed above are likely to belong to one species, then these others might all belong to a second. The type of pygidial pleural ribbing is quite like that of certain *Astroproetus* species, e.g. *A. asteroideus* (Begg,

1939), which also has a weakly convex preglabellar field, like Temple's 'cranidium type 2' (cf. Pl. 11, fig. 2b and Temple 1969, pl. 4, figs. 6, 14, 18, 19), although the incised glabellar furrows seen in smaller specimens (Temple 1969, pl. 4, fig. 6) have no counterparts in known *Astroproetus* species.

10. **Decoroproetus scrobiculatus** sp. nov.   Pl. 9, figs. 10–20; Pl. 10, figs. 1–8, Text-fig. 7

Name. Latin 'scrobiculus', meaning a little trench;
referring to the presence of impressed glabellar furrows.

1915   *Proetus gracilis* Barrande; Hede, p. 45, pl. 4, figs. 22–24.
1965   *Decoroproetus* sp.; Rickards, p. 548.
1967   *Decoroproetus*; Rickards, p. 230.

Type specimens. Holotype, NMW 71.6G.512a, internal mould of a cranidium, with incomplete counterpart external mould, Pl. 9, figs. 10a–c; paratypes: HUR ID/119, external mould of cranidium; HUR ID/241, external mould of free cheek; HUR ID/276, internal mould of hypostome; from Silurian, highest Wenlock (*ludensis* Zone) and basal Ludlow (*nilssoni-scanicus* Zone) bipartite limestone, small exposures near E side of Sedbergh–Kirkby Stephen road (A683), 1390 yd W of Bluecaster Hill and 4 miles NE of Sedbergh, NW Yorkshire (SD 6995 9698).

Material, horizons and localities. There are approximately 100 cranidia, free cheeks, hypostomes and pygidia belonging to this species from the type locality in the Rickards Collection at Hull University, mostly internal moulds. Detached exoskeletal parts, TCD 9610–16 from W of Glyn Farm (SJ 3035 1167), TCD 9617 from Weston Brook, N of Binweston (SJ 3005 0444), TCD 9618 from lane NW of Worthen (SJ 3242 0513), TCD 9630 from track N of Worthen (SJ 3283 0527), all from Trewern Brook Formation (Wenlock Series, highest *lundgreni* Zone and *nassa-dubius* Interregnum), Long Mountain, Welsh Borderland. Cranidia LPI LO2843t and LPI LO2844t and pygidium LPI LO2845t from Wenlock Series (*ludensis* Zone), Smedstorp, Scania, S Sweden.

Diagnosis. Glabella with 3 pairs of weakly impressed furrows; occipital ring without lobes; preglabellar field sigmoidal; anterior border broad, flattened, lateral border broadening markedly towards base of genal spine; irregular tropidial ridges on lateral parts of preglabellar field and on anterior part of free cheek; pygidial axis with 7–9 rather ill-defined rings, pleural areas with 5 or 6 pairs of ribs. Sculpture of fine striations.

Description. Cranidium with sagittal length rather greater than palpebral width. Glabella weakly inflated, tapering gently forwards to well-rounded frontal lobe. Lateral constriction hardly apparent on large specimens (e.g. Pl. 9, figs. 10, 15, 16), but quite marked on small ones (Pl. 10, figs. 1, 2, 4). Three pairs of glabellar furrows, all weakly incised. 1p: opposite anterior end of palpebral lobe, deeper than 2p and 3p, directed obliquely backwards at about 45° and has a small auxiliary impression associated (Pl. 9, fig. 16). 2p: almost opposite γ, running slightly backwards; 3p: isolated from axial furrow, runs nearly straight adaxially and is about same length as 2p.

Occipital furrow of comparable width and depth to axial furrows, arched weakly backwards sagittally. Occipital ring as wide (sag.) as preglabellar field and as wide, or a little wider (trans.) than glabella, and maintains more or less the same width (sag. and exsag.). No lateral occipital lobes, but a small median tubercle is present (e.g. Pl. 9, figs. 16, 17).

Preglabellar field between $\frac{1}{4}$ and $\frac{1}{5}$ length (sag.) of glabella, sigmoidal in longitudinal section (see Pl. 9, fig. 10b). On lateral parts of preglabellar field and on anterior part of free cheek are a series of irregular tropidial ridges (e.g. Pl. 9, figs. 15, 16). Section β–γ of anterior branch of facial suture diverges abaxially forwards from γ at 18°–22° in larger specimens, 34°–38° in very small specimens. γ close to axial furrow. On posterior branches ϵ and ξ fall on one continuous curve (Pl. 10, fig. 1), with suture reaching close to axial furrow.

Palpebral lobe posteriorly placed, subcrescentic, rather narrow and between $\frac{1}{2}$ and $\frac{1}{3}$ sagittal length of glabella. Eye large, crescentic, mounted on a distinct eye socle, whose lower margin diverges gently from the upper at either end. Field of free cheek narrow, gently convex.

Cephalic border furrow shallow but quite well defined. Anterior border broad, weakly inflated, lateral border flattens and widens towards the base of the genal spine (Pl. 9, fig. 13). Latter is broad based and blade-like, with shallow median furrow, which extends for its entire length, dividing it into a broader outer and a narrow inner band. Posterior border furrow narrow and deep, abruptly truncated at base of genal spine.

Cephalic doublure broad, with prominent, parallel terrace lines. Hypostome with strongly convex median body, with ill-defined, strongly backwardly directed median grooves. Anterior wing directed upwards at a high angle. Anterior border narrow, separated from median body by a distant anterior border furrow. Lateral and posterior border furrows wider and deeper than anterior, and convex lateral and posterior borders are wider than anterior. On posterior margin is pair of short, broad based spines.

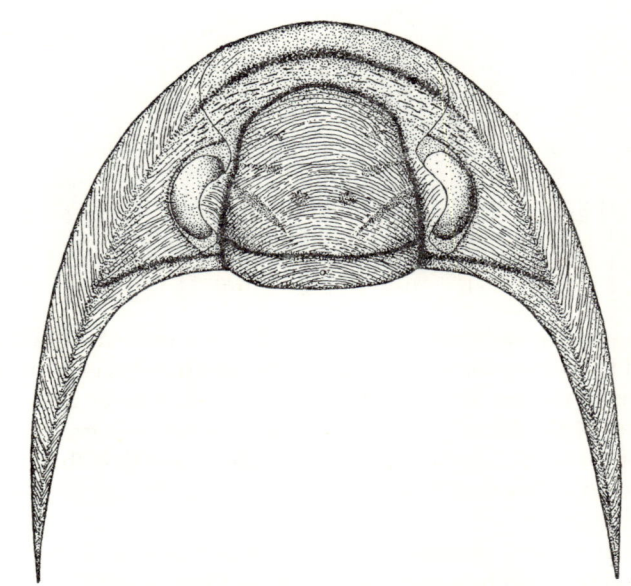

TEXT-FIG. 7. Reconstruction of the cephalon of *Decoroproetus scrobiculatus* sp. nov. For clarity, the striated sculpture has been somewhat exaggerated. Based on Pl. 9, figs. 10, 11, 13, 16. ×6 approx.

Number of thoracic segments unknown. Axis broad, anteriorly wider (trans.) than pleural area. Annulus slightly wider (sag.) than articulating half-ring. Pleura with deep pleural furrow, running very slightly backwards in abaxial direction. Adaxially posterior pleural band wider, abaxially anterior one is wider. Posterolateral corner of pleura distinctly angular.

Pygidium subparabolic. Axis anteriorly between $\frac{1}{4}$ and $\frac{1}{3}$ total pygidial width, and tapers rather strongly backwards, posterior end rather poorly defined. There is no distinct postaxial ridge. Seven to nine axial rings, defined by very shallow ring furrows. Pleural areas with five or six pairs of pleural ribs. Pleural furrows deep and pronounced, extending close to pygidial margin. Interpleural furrows indistinct, with only the first one or two pairs distinguishable. Pygidial doublure broad, weakly ventrally convex and with six or seven prominent terrace lines. Between these, and running parallel to them is a series of faint, secondary terrace lines (see Pl. 10, fig. 8).

Sculpture of fine, raised striations, covering entire exoskeleton.

Measurements.

Cranidia	A	A$_1$	A$_2$+A$_3$	A$_4$	K	δ–δ	
NMW 71.6G.512a (I)	9·9	6·0	2·4	1·5	5·3	6·3	HOLOTYPE
HUR ID/119 (E)	5·9	3·6	1·4	0·9	4·2	(5·5)	PARATYPE
TCD 9612 (E/I)	8·1	5·0	2·0	1·1	5·0	6·7	
TCD 9611 (I)	—	4·8	1·6	—	4·9	(6·0)	
TCD 9610 (I)	7·8	4·7	1·2	0·9	4·9	6·6	
LPI LO2843t (E/I)	—	4·1	—	0·9	4·6	—	
HUR ID/78 (E)	6·0	3·4	1·7	0·9	3·3	4·3	
LPI LO2844t (E/I)	(6·0)	3·6	(1·7)	0·7	(3·1)	—	
HUR ID/106 (I)	4·2	2·4	1·1	0·7	2·5	—	
HUR ID/91 (I)	3·8	2·4	0·9	0·5	2·5	(3·5)	
HUR ID/94 (I)	3·1	1·7	0·9	0·5	1·9	2·7	
HUR ID/143 (I)	—	1·2	0·7	—	(0·8)	(1·9)	
HUR ID/20b (I)	1·6	0·8	0·5	0·3	0·9	(1·1)	

Hypostomata	HL	HW$_1$	HW$_2$	HW$_3$	
HUR ID/276 (I)	4·4	4·2	2·9	3·0	PARATYPE
HUR ID/69 (I)	3·1	2·8	1·1	2·0	
HUR ID/273 (I)	3·4	(2·6)	2·3	2·6	

Pygidia	Z	Y	W	X
SM A38993 (I)	(6·0)	4·3	(12·6)	3·8
HUR ID/135 (I)	4·5	1·5	5·0	1·2
HUR ID/129 (I)	4·0	0·5	5·4	1·6
LPI LO2845t (I)	(1·9)	(1·3)	3·2	1·0
HUR ID/266 (I)	(1·2)	(1·0)	2·5	0·7
HUR ID/265 (I)	1·1	0·9	2·2	0·6

Remarks. Among the specimens from the type locality are two early holaspid cranidia and two early holaspid pygidia. The former show differences from mature specimens in that the preglabellar field is longer, 1p is deeper, the glabella is proportionately narrower and is distinctly constricted, and the anterior branches of the facial sutures are more strongly divergent. The latter are proportionately broader than in mature specimens. The cranidia are like that of *Decoroproetus? scanicus* from the Upper Ordovician of Scania, Sweden (see Owens 1973, fig. 8I) but mature specimens of this species have yet to be found.

   D. scrobiculatus is clearly closely related to the contemporaneous *D. decorus* (Barrande), which is found in Bohemia and Poland and has been refigured by Owens (1973, fig. 4C). The latter species has a much narrower cephalic border than *D. scrobiculatus*, a broader field of the free cheek and a more distinctly constricted glabella.

11. **Decoroproetus sp. 1**   Not figured

1970   Proetidae indet.; Temple, p. 61, pl. 18, figs. 1, 7, 8.

   Material, horizon and locality. Cephalon with incomplete thorax (SM A62519), cranidia, free cheeks (e.g. SM A62498) and pygidia (e.g. SM A62499) from Llandovery, Rhuddanian Stage, A$_{3–4}$, exposure beside Forestry track joining Meifod–Llanerfyl road (A 495), *c.* 3¼ miles ESE of Tanhouse Farm, Montgomeryshire (SJ 1135 1013).

   Remarks. This material, not assigned generically by Temple, is referable to *Decoroproetus* and somewhat resembles *D. scrobiculatus* although the lateral border of the free cheek does not widen towards the base of the genal spine.

12. **Decoroproetus sp. 2**   Pl. 10, figs. 9, 10

1938   *Proetus* sp. ind.; Whittard, p. 98, pl. 3, fig. 5.

   Material, horizons and localities. Cranidium, GSM 55472 (figured Whittard 1938, pl. 3, fig. 5) from a calcareous horizon in the Minsterley Formation [ZCM loc. 75] (Llandovery, Telychian Stage), Minsterley–Habberley lane, Shropshire (SJ 3803 0487); internal with counterpart external moulds of nearly complete specimen, SM A81488, from Llandovery, Fronian Stage, trackside exposure on E bank of stream 500 yd SSW of Lletty'r hyddod, near Llandovery, Carmarthenshire (SN 7143 2811).

Remarks. This species differs from *D. scrobiculatus* in having striations continuous across the preglabellar field and in having non-incised lateral glabellar furrows. The nearly complete specimen (Pl. 10, fig. 10) shows the lateral cephalic border flattened and widened near the base of the genal spine, like *D. scrobiculatus*.

13. **Decoroproetus sp. 3**   Pl. 10, figs. 11–17

Material, horizons and localities. GSM Da4907, 4910–13, free cheeks and pygidia from Wenlock Shale, cutting on N side of Martley to Clifton-on-Teme road, 400 yd S of Hillend farm, Worcestershire; NMW 72.18G.76, incomplete cranidium from Buildwas Beds, old brick pit, *c*. 500 yd SSE of Ticklerton Hall Farm, Shropshire (SO 4858 9044); GSM D4010–11, 4016, 4043, 4059, one nearly complete internal mould and cranidia and pygidia from Wenlock Shale, excavation 470 yd at 65° from Church Stretton church, Shropshire (SO 4565 4059).

Remarks. The better preserved material from Marltey and Ticklerton shows a striated surface sculpture and, as in *D. scrobiculatus*, the striations do not cross the preglabellar field. Principal differences from that species appear to be the even width of the lateral cephalic border and the shorter, more conical pygidial axis. More and better material is required before it will be possible to judge the affinities of this and the two foregoing species in more detail, and to find out how they are related to one another and to *D. scrobiculatus*. It is not impossible that one or more may turn out to be conspecific with the latter.

Genus **ASTROPROETUS** Begg, 1939

(= *Clypoproetus* Begg, 1939; *Sibiroproetus* Přibyl, 1970; ?*Enodiproetus* Přibyl, 1970).

Type species. Originally designated by Begg 1939, p. 375; *Proetus (Astroproetus) reedi* Begg, 1939, p. 375; from Upper Drummuck Group (Ashgill Series, Rawtheyan Stage), Lady Burn, near Girvan, Ayrshire.

Diagnosis. Glabella typically conical; glabellar furrows, when present, are weakly impressed; preglabellar field sigmoidal or weakly convex; no tropidium; lateral occipital lobes always present; thorax of 9–10 segments; pygidium with long, rather narrow axis with 7–8 rings; pleural areas with 5–6 pairs of ribs with weakly imbricate profile; exoskeleton smooth, except in some species with pits on the cheeks.

Remarks. Although *Astroproetus* superficially resembles *Decoroproetus*, the conical glabella, lateral occipital lobes, the gentle backward curve of the pygidial pleural furrows which are of nearly uniform depth along their length, and the non-striated exoskeleton combine to distinguish it from that genus. The two genera were further discussed by Owens (1973, p. 134).

The presence of lateral occipital lobes and the long, narrow pygidial axis in *Astroproetus* are characters comparable with *Warburgella* Reed, 1931, but the latter differs in two important aspects —in having the connective sutures of the rostral plate diverging backwards and in having deep 1p furrows. It is the lack of these latter characters which precludes the reference of *Astroproetus* to the Warburgellinae. The common characters of the two genera, however, suggest that *Astroproetus* might be associated with the ancestral stock of *Warburgella*. In the Ordovician, *Astroproetus* is known only from the Upper Drummuck Group in the Girvan area, but in the Silurian (Llandovery) it is known from the Girvan area, from S Wales, from Percé, Quebec, from the Oslo region, Norway, and from Siberia.

The type species of *Clypoproetus* Begg, 1939, *C. asteroideus* (Pl. 11, figs. 2–4), differs from *A. reedi* only in a few minor characters which are not considered to be of generic value. I thus agree with Whittington (1966a, p. 81) that *Clypoproetus* is a synonym of *Astroproetus*.

Přibyl (1970) has recently erected *Sibiroproetus* with the type species *Pseudoproetus bellus* Maksimova, 1962, from the Llandovery of the Wilujskoje Plateau, Siberia, and the subgenus *Unguliproetus (Enodiproetus)* with the type species *Proetus enodis* Maksimova, 1955, from the Llandovery of the Oldondo River, central Siberia. I have seen silicone rubber casts of the type material of both these species kindly made available by Professor Dr. H. K. Erben. Přibyl distinguished *Sibiroproetus* from *Astroproetus* by "the marked basal lobes of the glabella, the lack of lateral occipital lobes, a different path of the facial suture, the eye size and the other characters" [my translation] exhibited by the latter. To judge from these remarks, Přibyl was evidently misled by Begg's (1939, p. 375) incorrect statement that *Astroproetus* had well developed basal lobes, and by the *Treatise* in which illustration of this genus (Moore 1959, p. O396, fig. 301, 4) is very inaccurate. None of the characters by which Přibyl distinguished *Sibiroproetus* from *Astroproetus* is found in the latter, and there is no evident reason for retaining *Sibiroproetus*. All the characters of the type species—the glabellar outline, the well developed lateral occipital lobes and the type of pygidial pleural ribs—are typical of *Astroproetus*.

The pygidial pleural rib-structure of the type species of *Enodiproetus* does not suggest association with the proetine *Unguliproetus*, but it is very similar to that of *Astroproetus*, and if not actually congeneric, is certainly closely related.

Species	Glabellar outline	Lateral occipital lobes	Preglabellar field	Number of thoracic segments	Number of pygidial axial rings
reedi	conical	ovate	sigmoidal	10	7
asteroideus	"	"	weakly convex	9	7–8
interjectus	"	"	sigmoidal	?	?
scoticus	"	transversely elongated	weakly sigmoidal	?10	7–8
pseudolatifrons	elongate trapezoidal	"	weakly convex	?	?8

TABLE 5. Summary of diagnostic characters of *Astroproetus* species described herein.

1. **Astroproetus reedi** (Begg, 1939) Pl. 10, figs. 18–21; Pl. 11, fig. 1

1939 *Proetus (Astroproetus) reedi* Begg, p. 375, pl. 6, fig. 2.
1950 *Proetus fardenensis* Begg, p. 285, pl. 14, figs. 1, 2.
1966a *Astroproetus reedi* Begg; Whittington, p. 81, pl. 25, figs. 7, 10, 11.
1967 ?*Warburgella reedi* (Begg); Ormiston, p. 62.
1969 *Astroproetus reedi* Begg, 1939; Pillet, p. 73, pl. 4, fig. 9, pl. 6, fig. 26.

Holotype. HM A1082, Pl. 10, figs. 21a, b; a complete internal mould figured Begg 1939, pl. 6, fig. 2, refigured Whittington 1966a, pl. 25, figs. 7, 10, 11; from Upper Drummuck Group, Starfish Bed no. 2 (Ashgill, Rawtheyan Stage), Lady Burn, near Girvan, Ayrshire.

Material, horizons and localities. This species has been recorded only from the Upper Drummuck Group exposed in Lady Burn, near Girvan, Ayrshire. The material consists of several complete or nearly complete internal moulds—BM In41431, BM In21917 (both with counterpart external moulds) BM In43102, BM In40839, BM In40956, BM I16127 from Starfish Bed no. 1 and HM A3821 (holotype of *Proetus fardenensis*) from Starfish Bed no. 2.

Diagnosis. Glabella conical, lateral glabellar furrows not apparent; eye large, about $\frac{2}{3}$ sagittal length of glabella; preglabellar field sigmoidal in longitudinal section; lateral occipital lobes small, ovate; thorax of 10 segments; pygidial axis with 7 rings, pleural areas with 5 pairs of ribs.

Description. Cephalic outline parabolic, with rather narrow, weakly convex border, defined by shallow, but distinct anterior and lateral border furrows. Glabella conical, as wide (trans.) or a little wider than long (sag.), rather weakly inflated and with no apparent lateral furrows. Occipital furrow deeper and wider than axial furrow, arched forwards weakly laterally and sagittally. Anterior slope vertical, posterior slope about 45°. Occipital ring slightly wider (sag.) than anterior border, maintaining constant width laterally (exsag.), and as wide (trans.) as glabella. Small, ovate, prominent lateral occipital lobes present.

Preglabellar field about $\frac{1}{4}$ the sagittal length of glabella, and is sigmoidal in profile. Section $\beta-\gamma$ of anterior branch of facial suture diverges abaxially forwards at 32°–38° from γ, which is close to axial furrow. Posterior branches with ϵ and ξ as independent angles, the former close to posterolateral corner of glabella, latter opposite lateral occipital lobe. Stretch $\epsilon-\xi$ close to and parallel with axial furrow.

Palpebral lobe large, crescentic, close to glabella, backwardly placed and about $\frac{1}{3}$ the glabellar sagittal length, in transverse section inclined at about 45° from axial furrow, flattening abaxially. Eye prominent, crescentic and about $\frac{2}{3}$ sagittal length of glabella. No distinct eye socle. Field of free cheek moderately convex. Posterior border furrow narrow and deep, truncated at base of genal spine. Posterior border about same width as anterior, widening slightly abaxially. Genal spine broad-based, apparently with wide, shallow median groove, and extends backwards as far as sixth thoracic segment.

Cephalic doublure narrow, strongly ventrally convex with prominent, parallel terrace lines. Rostral plate (see Pl. 10, figs. 19, 21) is trapezoidal, transversely elongated, with connective sutures converging backwards.

Thorax of ten segments. Axis tapers evenly backwards so that last ring is about $\frac{1}{2}$ width (trans.) of first, and in transverse section is quite strongly convex. Each ring arched forward sagittally and laterally. Pleurae gently declined at fulcrum. Each pleura has narrow, distinct, pleural furrow, dividing pleura into anterior and posterior bands of approximately equal width (exsag.), and is truncated by posterior edge of articulating facet. Posterolateral corner of pleura angular.

Pygidium subparabolic, without border, with strongly longitudinally convex, narrow axis. Axis occupies about $\frac{1}{4}$ pygidial width (trans.), tapers gently backwards and consists of seven rings, defined by shallow ring furrows. Narrow postaxial ridge present. Pleural areas gently convex, with five pairs of pleural ribs which curve gently backwards and widen a little abaxially. Interpleural furrows inconspicuous, almost parallel with pleural. Both pleural and interpleural furrows die out before reaching margin. Anterior and posterior bands of nearly equal width (exsag.) and convexity. Pygidial doublure wider and less convex ventrally than cephalic, and is likewise ornamented with fine, parallel terrace lines.

One specimen (Pl. 10, fig. 19) with a damaged glabella shows the external mould of the hypostome. This has strongly convex median body, rather deep lateral border furrows, prominent anterior wings and strongly forwardly curved anterior margin. Shoulder well developed, extending abaxially as far as anterior wing. Posterior margin apparently transverse.

Entire dorsal exoskeleton smooth.

Measurements.

Cranidia	A	A_1	A_2+A_3	A_4	K	$\delta-\delta$	
HM A1082 (I)	5·5	3·3	1·4	0·8	3·4	5·1	HOLOTYPE
HM A3821 (I)	5·6	3·4	1·6	0·6	4·9	(6·4)	Holotype of *fardenensis*
BM In43102 (I)	5·5	3·7	1·2	0·6	(4·2)	(4·9)	
BM In21917 (I)	5·4	3·5	1·1	0·8	4·5	—	
BM In40839 (I)	(5·4)	3·9	(1·0)	0·5	4·1	(4·9)	
BM In41431 (I)	4·4	2·6	1·4	0·4	(3·2)	—	

Pygidia	Z	Y	W	X	
HM A1082 (I)	4·2	3·4	6·1	1·9	HOLOTYPE
BM In40839 (I)	4·7	3·9	8·4	2·2	
BM In21917 (I)	3·4	2·7	8·0	1·8	
BM In41431 (I)	2·9	2·4	6·8	1·8	

Remarks. Begg (1939, p. 375) stated that this species had 'pronounced basal lobes', but their apparent presence on the holotype seems to be a result of crushing, and the feature is not seen on any of the other specimens attributable to this species. Lateral glabellar furrows are not seen on any specimen, but their apparent absence may be due to bad preservation. The external mould of this species (Pl. 10, fig. 18b) clearly demonstrates the presence of distinct lateral occipital lobes, whose presence Whittington (1966a, p. 81) suspected from examining internal moulds.

Proetus fardenensis Begg, 1950 can be regarded as a synonym of *A. reedi*, although the holotype of that species shows a proportionately wider glabella than the holotype of *reedi*. This seems to have been brought about, at least in part, by compression, and there are no other features of *fardenensis* which warrant its distinction from *reedi*.

2. **Astroproetus asteroideus** (Begg, 1939)   Pl. 11, figs. 2–4

1939   *Proetus (Clypoproetus) asteroideus* Begg, p. 374, pl. 6, fig. 1.
1966a *Astroproetus asteroideus* (Begg); Whittington, p. 81, pl. 25, figs. 8, 9, 13.
1969   *Clypoproetus asteroideus* Begg; Pillet, p. 73, pl. 4, fig. 10; pl. 6, fig. 27.

Holotype. HM A1080, Pl. 11, figs. 2a, b; a nearly complete internal mould, figured Begg 1939, pl. 6, fig. 1, refigured Whittington 1966a, pl. 25, figs. 8, 9, 13; from Upper Drummuck Group, Starfish Bed no. 2 (Ashgill, Rawtheyan Stage), Lady Burn, near Girvan, Ayrshire.

Material, horizons and localities. This species is known only from the Upper Drummuck Group, exposed in Lady Burn, near Girvan, Ayrshire. All the specimens are complete or partially complete internal moulds—BM In21920, BM In21945, BM In40958, BM In40969, BM In42718, BM In42740, BM In46895, from Starfish Bed no. 1, HM A1081 from Starfish Bed no. 2.

Diagnosis. Glabella conical, with weakly impressed furrows; preglabellar field weakly convex in profile; lateral occipital lobes small, ovate; thorax of 9 segments; pygidial axis with 7–8 rings, pleural areas with ?5 pairs of ribs.

Comparison. This species is very similar to *A. reedi*, differing only in the convex preglabellar field and having 9 rather than 10 thoracic segments. One specimen (Pl. 11, fig. 3) shows evidence of three pairs of weakly impressed, gently backwardly curved lateral glabellar furrows, a feature not seen on any of the material of *A. reedi*.

Measurements.

Cranidia	A	A_1	A_2+A_3	A_4	K	δ–δ	
HM A1080 (I)	2·4	1·6	0·5	0·3	1·5	(0·9)	HOLOTYPE
BM In46895 (I)	(8·7)	6·0	1·8	(0·9)	2·3	5·5	
BM In21945 (I)	(8·0)	5·7	(1·5)	0·8	1·9	5·0	
HM A1081 (I)	(4·9)	2·7	1·7	—	2·4	(2·8)	
BM In42740 (I)	2·4	1·4	0·8	0·2	1·2	(1·4)	

Pygidia	Z	Y	W	X	
HM A1080 (I)	1·4	1·2	2·4	0·8	HOLOTYPE
BM In46895 (I)	5·7	5·0	8·6	3·2	

Remarks. *A. asteroideus* has been identified in the Gray Collection in the British Museum (Natural History), and most of the specimens are better preserved than the holotype and clearly demonstrate that this species has nine thoracic segments. The locality and horizons of *A. asteroideus* are the same as those of *A. reedi*, and the differences are small, and could be of sexual rather than specific significance.

3. **Astroproetus scoticus** (Reed, 1941)   Pl. 11, figs. 5–10

1904   *Proetus pseudolatifrons* Reed, p. 78 *partim*, pl. 11, fig. 9 (*non* figs. 7, 8).
1904   *Proetus* cf. *obconicus* Lindström; Reed, p. 81, pl. 11, fig. 12.
1904   *Proetus* sp. ind.; Reed, p. 81, pl. 11, fig. 13.
1941   *Proetus scoticus* Reed, p. 271, pl. 5, fig. 3.
1960   *Proetus scoticus* Reed; Kielan, p. 183.
1970   *Sibiroproetus*? sp. n.; Přibyl, p. 108.

Holotype. By monotypy; HM A1109, Pl. 11, fig. 7; internal mould of cranidium, figured Reed

1941, pl. 5, fig. 3; from Mulloch Hill Sandstone (Llandovery Series, Rhuddanian Stage), old quarry *c*. 1000 yd ESE of Craigens, near Girvan, Ayrshire (NS 2594 0424).

Material, horizons and localities. This species has been recorded only from the early Llandovery (Rhuddanian Stage) of the Girvan district. BM In21961, cranidium and BM In21950, pygidium, both from type locality. BM In42682, and BM In58601, both almost complete moulds and BM In21938, BM In21962, BM In42685, cranidia from Mulloch Hill; BM In21959, cranidium from Woodland Point.

Diagnosis. Broad conical glabella with 3 pairs of weakly impressed lateral furrows; lateral occipital lobes transversely elongated; preglabellar field straight or weakly sigmoidal; eye rather small; pygidial axis with 7–8 rings, broad pleural areas with 5–6 pairs of ribs.

Description. Cephalon with moderately wide, weakly convex border defined by shallow but distinct anterior and lateral border furrows. Glabella broadly conical, varying from being wider (trans.) than long (sag.) to being a little longer than wide, and is rather weakly inflated. Three pairs of weak lateral glabellar furrows seen on some specimens (e.g. Pl. 11, fig. 6). 1p: opposite anterior part of palpebral lobe, directed rather weakly backwards; 2p: nearly opposite γ; 3p: close to anterolateral corner of glabella.

Occipital furrow deep and broad, describing a gentle, posteriorly convex curve, with median section, between adaxial ends of lateral occipital lobes, nearly transverse in holotype. Occipital ring as wide (trans.) as glabella with prominent, transversely elongated lateral occipital lobes. Median tubercle not seen on any specimen.

Preglabellar field straight or weakly sigmoidal in profile, about as long (sag.) as anterior border. Section β–γ of anterior branch of facial suture diverges abaxially forwards at 18°–34° from γ, which is a wide angle a little distance out from axial furrow. Posterior branches with ϵ and ξ as independent angles, the intervening stretch parallel with and close to axial furrow. Palpebral lobe crescentic, approximately $\frac{1}{3}$ sagittal length of glabella. Eye rather small, without distinct eye socle.

Field of free cheek broad, weakly convex. Posterior border furrow deeper than anterior and lateral, abruptly truncated at base of rapidly tapering genal spine, which extends backwards as far as fifth or sixth thoracic segment. Cephalic doublure as wide as border, ventrally convex with distinct, parallel terrace lines.

The most complete specimen (Pl. 11, fig. 5) has parts of ten thoracic segments preserved, which is assumed to be the full complement. Axis rather narrow, slightly wider (trans.) than pleurae at anterior end, but narrower posteriorly. Each ring weakly convex and backwardly inclined in lateral profile (Pl. 11, fig. 5b). Pleura with deep pleural furrow, dividing it into a narrower anterior band and a wider posterior band. Pleural furrow truncated close to abaxial end of pleura, whose posterolateral corner is bluntly angular.

Pygidium subparabolic, axis anteriorly about $\frac{1}{4}$ total pygidial width, gently tapering, composed of seven to eight rings, defined by shallow ring furrows. Broad pleural areas with five or six pairs of pleural ribs, which curve gently backwards, widening slightly abaxially. Both pleural and interpleural furrows extend close to margin, the former being considerably deeper and more conspicuous than the latter. Pygidial doublure weakly ventrally convex, with distinct, parallel terrace lines.

As far as can be judged from the available specimens, the external surface is smooth.

Measurements.

Cranidia	A	A_1	A_2+A_3	A_4	K	δ–δ	
HM A1109 (I)	8·6	5·3	2·2	1·1	7·3	(10·2)	HOLOTYPE
BM In42682 (I)	(13·4)	9·4	(2·8)	1·2	7·8	(10·4)	
BM In58601 (I)	—	8·7	—	(1·3)	8·3	(9·8)	
BM In21959 (E/I)	9·1	6·4	1·4	1·3	5·6	(6·7)	

Pygidia	Z	Y	W	X
BM In42682 (I)	(10·9)	8·6	16·9	4·4
BM In21950 (I)	10·0	8·3	(15·7)	4·1
BM In58601 (I)	9·2	7·6	18·2	4·3

Remarks. This species differs from *A. reedi* in having a proportionately broader glabella, a shorter (sag.) preglabellar field, a smaller eye, transversely elongated rather than rounded lateral occipital lobes and broader pygidial pleural areas.

From Llandovery (Fronian Stage, C_1) beds of the Llandovery district Dr. P. D. Lane has collected a cranidium and a pygidium (Pl. 11, figs. 11, 12) of *Astroproetus*; these specimens are similar to *A. scoticus* (see Pl. 11, figs. 5–10), differing in the narrower anterior border, larger palpebral lobe and smaller occipital lobe. A separate species is probably represented, but until more material is available these specimens are referred to as *A. aff. scoticus*.

4. Astroproetus interjectus (Reed, 1935)   Pl. 11, figs. 13–15, Pl. 12, fig. 1

1904   *Proetus latifrons* (M'Coy); Reed, p. 76, pl. 11, figs. 4, 4a.
1935   *Proetus interjectus* Reed, p. 40, pl. 3, fig. 23.
1950   *Proetus subtriangularis* Begg, p. 287, pl. 14, fig. 6.

Holotype. By monotypy; HM A1031, Pl. 11, figs. 14a, b; small internal mould of cranidium, figured Reed 1935, pl. 3, fig. 23; from Saugh Hill Sandstones (Llandovery, Idwian Stage), Newlands, near Girvan, Ayrshire (NS 277 044).

Material, horizons and localities. This species is known only from the type locality, represented by BM In21934, BM In42699, HM A3822 (holotype of *subtriangularis*), HM A5022–23, all internal moulds of cranidia.

Diagnosis. Glabella conical; lateral occipital lobes small, ovate; preglabellar field sigmoidal; weak lateral glabellar furrows developed.

Description. Only cranidia of this species are known. The material is all indifferently preserved, so only a brief description is given here: glabella conical, varying from being approximately as long (sag.) as wide (trans.) to being somewhat longer than wide. Weakly impressed lateral glabellar furrows; preglabellar field sigmoidal, nearly ⅓ the length (sag.) of glabella. Anterior border furrow shallow, but distinct; anterior border rather narrow, upturned. Lateral occipital lobes small, ovate.

Measurements.

Cranidia	A	A_1	A_2+A_3	A_4	K	δ–δ	
HM A1031 (I)	2·1	1·3	0·5	0·3	1·4	—	HOLOTYPE
BM In21934 (I)	6·1	4·1	1·1	0·9	4·2	—	
BM In42699 (I)	5·4	3·2	1·3	0·9	3·7	—	
HM A3822 (I)	5·2	3·3	1·2	0·7	2·8	—	Holotype of *subtriangularis*

Remarks. The holotype cranidium of *Proetus subtriangularis* has a more elongated glabella than that of *interjectus*, but otherwise the two specimens are similar. As both are from the same horizon and locality, it is likely that the differences in glabellar proportions are due to variation, and hence *subtriangularis* is considered to be a synonym of *interjectus*. *A. interjectus* is similar to *A. reedi*, differing only in possessing weak lateral glabellar furrows and in apparently having less distinct lateral occipital lobes, but better material and information on other parts of the exoskeleton of *interjectus* are required for fuller comparison, and to judge whether two separate species really are represented.

5. Astroproetus pseudolatifrons (Reed, 1904)   Pl. 12, figs. 2–5

1904   *Proetus pseudolatifrons* Reed, p. 78 *partim*, pl. 11, figs. 7, 8 (*non* fig. 9).
1931   *Proetus (Prionopeltis?) pseudolatifrons* Reed; Reed, p. 14 *partim*.

Type specimens. From Reed's two syntypes, BM In21946, an internal mould of a cranidium, figured Reed 1904, pl. 11, fig. 7, refigured herein Pl. 12, figs. 4a, b, is selected as the lectotype. BM In21946, an internal mould of a pygidium on the same piece of rock as the lectotype, figured Reed 1904, pl. 11, fig. 8, refigured herein Pl. 12, fig. 2, is the paralectotype; from Camregan Group (Llandovery Series, Fronian Stage), Camregan Wood, near Girvan, Ayrshire.

Material, horizon and locality. This species has been recorded only from the type locality, and besides the type specimens there is one free cheek, BM In21949 and one pygidium, BM In21947.

Diagnosis. Glabella of elongate trapezoidal outline, weakly constricted laterally; preglabellar field weakly convex; lateral occipital lobes transversely elongated; pygidial axis weakly longitudinally convex with ?8 rings; pleural areas with ?6 pairs of ribs.

Description. Glabella of elongate trapezoidal outline, a little longer (sag.) than wide (trans.), weakly inflated and slightly constricted a short distance in front of γ. No lateral glabellar furrows seen on single available cranidium. Occipital furrow rather shallow, median section nearly transverse, lateral ends flexed forwards. Occipital ring about as wide (sag.) as preglabellar field, and as wide (trans.) as glabella. Weak, transversely elongated, lateral lobes present, though not conspicuous.

Preglabellar field distinctly broader (sag.) than anterior border, and is weakly convex in profile. Anterior border furrow distinct but shallow, anterior border narrow, weakly convex. Section β–γ of anterior branch of facial suture diverges abaxially forwards at 32° from γ, which is close to axial furrow. Posterior branches with ϵ and ξ widely separated, the long intervening stretch diverging weakly from axial furrow towards posterior. Palpebral lobe and eye not preserved.

Field of free cheek broad, lateral border and lateral border furrow like anterior. Posterior border furrow broad and shallow, no deeper than lateral. Genal spine with rather narrow base.

Pygidium broad, subparabolic with gently tapering axis which is about $\frac{1}{4}$ total pygidial width anteriorly and is composed of ?eight ill-defined rings. Pleural areas with ?six pairs of ribs, pleural furrows shallow, apparently extending close to margin, interpleural furrows not seen on available material.

Measurements.

Cranidia	A	A$_1$	A$_2$+A$_3$	A$_4$	K	δ–δ	
BM In21946 (I)	10·2	6·7	2·2	1·3	5·8	—	LECTOTYPE

Pygidia	Z	Y	W	X		
BM In21946 (I)	(7·3)	6·2	(12·0)	2·6	PARALECTOTYPE	
BM In21947 (I)	—	(6·1)	14·4	3·6		

Remarks. Reed (1904, pp. 78–79), noticing Salter's misinterpretation of McCoy's species *Proetus latifrons*, originally intended *pseudolatifrons* to incorporate Ludlovian specimens figured and described by Salter (1848, p. 337, pl. 6, figs. 1, 1a) as *P. latifrons*, as well as the Llandovery specimens from Girvan. Later Reed (1931, p. 14) evidently restricted *pseudolatifrons* to the Girvan material, and clearly stated that he had based the species on it, although he had quoted Salter's description of the Ludlow specimens as his original description of the species. The Ludlow specimens, however, belong to an entirely different form, *Proetus* (*Lacunoporaspis*) *obconicus* Lindström, and are hence not directly comparable with *A. pseudolatifrons*.

A. pseudolatifrons is distinguished from other *Astroproetus* species particularly by its elongated trapezoidal glabella with weak constriction, but is otherwise similar to *A. scoticus*.

6. Astroproetus aff. pseudolatifrons (Reed, 1904)   Pl. 12, figs. 6, 7

Material, horizon and localities. SM A38902, internal mould of cranidium, from Browgill Beds (Llandovery, *Phacops elegans* Limestone), 410 yd S of spring, Copple Bank Wood, Crummackdale, near Austwick, W Riding of Yorkshire; SM A38833, internal mould of pygidium, from same horizon as above, loose boulder in Crummackdale, 1000 yd SE of Crummack Farm.

Description. Cranidium with pyriform glabella, distinctly constricted opposite γ, with no furrows indicated and slightly longer (sag.) than wide (trans.). Preglabellar field about $\frac{1}{3}$ glabellar length (sag.), gently convex in profile; occipital ring apparently narrowing laterally; palpebral lobe large, crescentic, backwardly placed. Pygidium: about twice as wide (trans.) as long (sag.); axis short, conical, with at least four rings; pleural ribs not preserved; doublure very broad with six subparallel terrace lines, bunched together behind axis.

Remarks. The cranidium and pygidium described above are apparently the only proetids from the *Phacops elegans* Limestone, and are considered to belong to one species. The cranidium shows some similarity to *Astroproetus pseudolatifrons* in glabellar outline and in its weakly convex pre-glabellar field, although the pygidium, with its rather broad, rapidly tapering axis is not very similar to that species, and its doublure is much wider. The form represented by the present material, is assigned to *Astroproetus*, showing some affinities with *A. pseudolatifrons* and, like the latter, differing from other *Astroproetus* species in glabellar outline and in the longitudinal section of the preglabellar field.

Genus **PARAPROETUS** Přibyl, 1964

Type species. Originally designated by Přibyl 1964, p. 42; *Proetus girvanensis* Nicholson & Etheridge, 1879, p. 169; from Upper Drummuck Beds (Ashgill Series, Rawtheyan Stage) of Thraive Glen, near Girvan, Ayrshire.

Diagnosis. Glabella with weakly incised furrows; occipital ring without lobes; preglabellar field very short (sag.), no wider than anterior border; eye small and forwardly placed; no tropidium; thorax of 10 segments; pygidium subparabolic, axis with 5 or 6 rings, ring furrows well defined; pleural areas with 4 or 5 pairs of pleural ribs with imbricate profile; sculpture of fine striations, locally interspersed with granules.

Remarks. The imbricate pygidial pleural ribs and the lack of the preannulus combine to indicate assignment of this genus to the Tropidocoryphinae. There is superficial resemblance to the proetine *Ascetopeltis* Owens, 1973, but the latter possesses the preannulus and the structure of the pygidial pleural ribs (e.g. Pl. 5, figs. 4, 5) is similar to that found in *Proetus*. Přibyl (1964, p. 44) considered *Paraproetus* to be closely related to *Proetus*, but its lack of typical proetine characters mentioned above precludes such a relationship. *Paraproetus* appears, rather, to be a small late Ordovician offshoot from *Decoroproetus* or a similar genus, and no undoubted representatives are known outside the late Rawtheyan and Hirnantian Stages of the Ashgill of the British Isles and southern Poland.

Paraproetus girvanensis (Nicholson & Etheridge, 1879)   Pl. 12, figs. 8–14; Pl. 13, figs. 1–4

1879 *Proetus girvanensis* Nicholson & Etheridge, p. 169, pl. 12, figs. 7–10.
1879 *Proetus procerus* Nicholson & Etheridge, p. 174, pl. 12, fig. 11.
1904 *Proetus girvanensis* Nicholson & Etheridge; Reed, p. 74, pl. 11, figs. 1–3.
1904 *Proetus procerus* Nicholson & Etheridge; Reed, p. 77, pl. 11, figs. 5, 6, 6a.
1916 *Proetus girvanensis* Nicholson & Etheridge; Marr, p. 200.
1931 *Proetus girvanensis* Nicholson & Etheridge; Reed, p. 14.
1943 *Proetus girvanensis* Nicholson & Etheridge; Begg, p. 57, pl. 2, figs. 7, 7a.
1945 *Proetus scobiei* Begg *in* Begg & Reed, p. 261, pl. 1, figs. 1, 2.
1956 *Proetus* sp., Harper, p. 391.
1960 *Proetus procerus* Nicholson & Etheridge; Kielan, p. 69.
1961 *Proetus procerus* Nicholson & Etheridge; Whittard, p. 187.
1964 *Paraproetus girvanensis* (Nicholson & Etheridge); Přibyl, pp. 42, 44.
1964 *Paraproetus procerus* (Nicholson & Etheridge); Přibyl, p. 44.
1969 *Paraproetus girvanensis* (Nicholson & Etheridge); Pillet, p. 74, pl. 3, fig. 7; pl. 6, fig. 21.

Lectotype. Selected by Přibyl 1964, p. 42; BM In21926, Pl. 12, fig. 8; a complete partially exfoliated external mould, figured Nicholson & Etheridge 1879, pl. 12, fig. 10. The other syntypes of Nicholson & Etheridge, GSM 35612, BM In21925 are paralectotypes. All from Upper Drummuck Group (Ashgill Series, Rawtheyan Stage), Thraive Glen, near Girvan, Ayrshire.

Material, horizons and localities. *Paraproetus girvanensis* is one of the commonest trilobites in the Upper Drummuck Group, and there is a large number of specimens, many complete, from the type locality and nearby in Thraive Glen. It is well represented in the Gray and McPherson collections in the British Museum (Natural History), e.g. BM In21914, In21915, (figured Reed 1904, pl. 11, figs. 2 and 3 respectively), In21942 (holotype of *procerus*), In21943 (figured Reed 1904, pl. 11, fig. 6), In21923–4, In40945, In41627, In46890–93, In46898, In 46906–7, I16050, I16124, I16127; and in the Begg Collection in the Hunterian Museum, Glasgow, e.g. HM A3689 (holotype of *scobiei*), A731. The species has also been recorded in small numbers from the topmost Rawtheyan *D. mucronata* Beds of the Lake District (SM A43154–56 from NNE of Sheepfold SW of Torver Beck (SD 2756 9613); SM A36002, SM A36010, SM A36028, SM A36037, SM A36049, SM A36051,

SM A36062, SM A36077 from waterfall on right side of Ash Gill Beck, SW of Ashgill Quarry (SD 2681 9535); GSM DJ409 from Appletreeworth Beck, 4 miles SW of Coniston, Lancashire (SD 2488 9309), and from probable Rawtheyan mudstones in streamside exposure in Afon Dwyfach, *c.* 1 mile N of Llanystumdwy, Caernarvonshire (NMW 72.22G.1) and (BM It8856–58) in temporary exposure in Dynanau farmyard, *c.* ¾ mile NNE of Llanystumdwy (SH 481 396).

Diagnosis. Glabella weakly constricted laterally or not constricted; eye about ¼ length of glabella, lower margin of eye socle diverging from upper quite strongly at either end, not incised; sculpture of fine striations, with a row of granules on posterior margins of occipital ring, thoracic and pygidial axial rings.

Description. Cephalon subparabolic with a narrow, convex border defined clearly by shallow anterior and lateral border furrows. Glabella subquadrate, commonly wider (trans.) than long (sag.), but width and length may be equal. Weak lateral constriction may be apparent (e.g. Pl. 12, fig. 10), but many specimens (e.g. Pl. 12, figs. 8, 9, 13) show little sign of it. Inflation weak, frontal margin well rounded. Three pairs of furrows; 1p: opposite middle of palpebral lobe, running obliquely backwards at about 45°, not reaching occipital furrow. Faint auxiliary impression associated (e.g. (Pl. 12, fig. 13a). 2p: opposite γ, extends a little over half-way towards sagittal line, runs weakly backwards. 3p: close to anterolateral corner of glabella, shorter than 2p and running weakly forwards.

Occipital furrow deeper and broader than axial furrows, flexed forwards sagittally and laterally. Deepest and widest about half-way between sagittal line and axial furrow. Anterior slope vertical, posterior slope inclined at about 45°. Occipital ring about as wide as preglabellar area, but exact proportions are rather variable (see measurements below). Transversely is as wide or a little wider than glabella. No lateral occipital lobes, and anterolateral corners of occipital ring are fused with posterolateral corners of glabella (e.g. Pl. 12, fig. 9b).

Preglabellar field short (sag.) less than ⅙ sagittal length of glabella, and in profile is weakly convex. Anterior branches of facial sutures vary from being nearly parallel (e.g. Pl. 12, fig. 13) to diverging abaxially forwards from γ at 24° (e.g. Pl. 12, fig. 14). γ close to axial furrow. Posterior branches with ϵ and ξ as widely separated, distinct angles, the intervening stretch close to and more or less parallel with axial furrow (see Pl. 12, fig. 13a).

Palpebral lobe small, semi-oval and well forwards, in transverse section inclined at about 55° from axial furrow. Eye small, about ¼ sagittal length of glabella. Eye-socle prominent, with lower margin diverging markedly from upper at either end, and not defined by an incised furrow. Field of free cheek broad, convex and with a distinct flexure running parallel with lateral and posterior margins (e.g. Pl. 12, fig. 13), and outside the flexure is strongly declined. Genal spine narrow, extending backwards as far as third thoracic segment, median groove only apparent at anterior end. Posterior border furrow broad and shallow.

Hypostome with prominent anterior wings, strongly convex median body and apparently rather deep median furrow. Posterior margin seems to lack spines.

Thorax of ten segments. Axis tapers gently backwards so that the last ring is about ⅔ width (trans.) of first. Annulus wider (sag.) than articulating half ring, and the two being divided by a deep articulating furrow. Doublure of axial ring dorsally convex, with strong, parallel, transverse terrace lines. Pleura with narrow, distinct pleural furrow, dividing it into narrow anterior and wide posterior pleural band. Pleural furrow truncated abaxially by posterior margin of articulating facet. Posterolateral corner of pleura distinctly angular.

Pygidium subparabolic, approximately twice as wide (trans.) as long. Axis anteriorly 25–33% of total pygidial width, with five or six rings differentiated by distinct ring furrows. Poorly defined postaxial ridge present. Pleural areas with four to five pairs of ribs which curve gently backwards. Pleural furrows shallow and sharp, maintaining more or less same depth along their length, reaching close to margin. Interpleural furrows weak, inconspicuous on external surface (e.g. Pl. 12, figs. 9b, 13d) but more distinct on internal moulds (e.g. Pl. 12, figs. 10, 11). Anterior and posterior pleural bands of more or less equal width.

Sculpture of fine striations, apparently confined to axial parts of exoskeleton, with distinct rows of granules on posterior margins of occipital ring, thoracic and pygidial axial rings (e.g. Pl. 12, fig. 9b).

Measurements.

Cranidia	A	A_1	A_2+A_3	A_4	K	$\delta-\delta$	
BM In21926 (E/I)	8·0	5·8	1·0	1·2	6·0	6·5	LECTOTYPE
GSM 35612 (E/I)	10·3	7·0	1·7	1·6	7·8	(9·4)	PARALECTOTYPE
BM In21914 (I)	(9·2)	6·4	(1·1)	1·7	6·8	8·3	
BM In46906 (I)	8·0	5·7	1·5	0·8	5·1	(6·0)	
HM A3689 (E/I)	7·3	4·6	1·4	(1·3)	6·1	6·5	Holotype of *scobiei*
BM In46907 (I)	6·8	4·8	1·1	0·9	4·3	(5·6)	
BM In21915 (I)	6·6	5·1	(0·8)	(0·7)	5·6	6·5	
BM It8858 (I)	(6·5)	3·8	—	1·0	(5·4)	(5·7)	
BM It8856 (I)	5·8	4·0	0·8	1·0	4·7	5·1	
GSM 32988 (E/I)	(5·8)	3·9	0·8	—	5·2	5·3	
BM In21942 (I)	5·0	3·3	0·9	0·8	3·3	(3·9)	Holotype of *procerus*
BM In40844 (I)	(4·9)	3·5	1·0	—	4·6	(5·5)	
BM In21943 (I)	(4·1)	3·1	0·7	(0·3)	2·9	3·4	
BM In40945 (I)	3·0	2·0	0·4	0·1	2·0	2·3	

Pygidia	Z	Y	W	X	
BM In21926 (E/I)	4·0	2·9	8·8	2·5	LECTOTYPE
GSM 35612 (E/I)	7·9	6·0	13·8	4·5	PARALECTOTYPE
BM In21914 (E)	5·2	4·3	10·3	3·3	
BM In46906 (I)	4·5	3·8	8·6	2·7	
BM In46907 (I)	3·8	2·8	7·3	2·0	
BM In21915 (I)	(3·6)	3·0	8·8	3·1	PARALECTOTYPE
BM In21942 (I)	2·3	2·0	5·2	1·2	Holotype of *procerus*
BM In21943 (I)	1·9	1·5	4·7	1·2	
BM In40945 (I)	1·5	1·1	2·8	1·0	

Remarks. Three species, *girvanensis* Nicholson & Etheridge, 1879, *procerus* Nicholson & Etheridge, 1879, and *scobiei* Begg, 1945, have been described from the Upper Drummuck Group of Thraive Glen, and all are referable to *Paraproetus*. The differences between the type specimens of these three species are in proportion only, and one of the types, that of *scobiei* (Pl. 12, fig. 14), is badly crushed. Examination of a large number of specimens of *Paraproetus* from the Upper Drummuck Group and measurement of glabellar parameters has shown that the types of the three species can be placed within a continuously variable series, as Text-fig. 8 shows, and only one species is represented in the sample. Begg (1939, p. 373 and 1943, p. 57) suggested that *girvanensis* and *procerus* are conspecific, but no revision of these species has been undertaken until now.

Kielan (1960, pp. 181–2, pl. 3, figs. 1, 2; pl. 21, figs. 1, 2) described and figured "*Proetus*" sp. a and "*P.*" sp. b from Ashgill strata (*Dalmanitina mucronata* Zone) of Zalesie, Holy Cross Mountains, Poland. Comparison of her figures with specimens of *P. girvanensis* shows several common characters—a very small preglabellar field, weakly impressed glabellar furrows, the anterolateral corners of the occipital ring running into the posterolateral corners of the glabella, small palpebral lobes, well-defined pygidial axial rings and gently backwardly-curving pleural ribs with narrow pleural furrows and inconspicuous interpleural furrows. These characters indicate assignment of the Polish specimens to *Paraproetus*, but the different sculpture (fine striations interspersed with granules) suggests that the species represented is probably distinct from *girvanensis*.

Subfamily WARBURGELLINAE subfam. nov.

Genera and subgenera. Warburgella (*Warburgella*) Reed, 1931; *Warburgella* (*Tetinia*) Chlupáč, 1971; *Prantlia* Přibyl, 1946; *Tropidocare* Chlupáč, 1971; possible member: *Koneprusites* Přibyl, 1964.

Diagnosis. Glabella with deep 1p furrows; preglabellar field always present; tropidium or tropidial ridges may be present; rostral plate with connective sutures diverging backwards; thorax of 8–10 segments, without preannulus; pygidium with narrow axis composed of 7–14

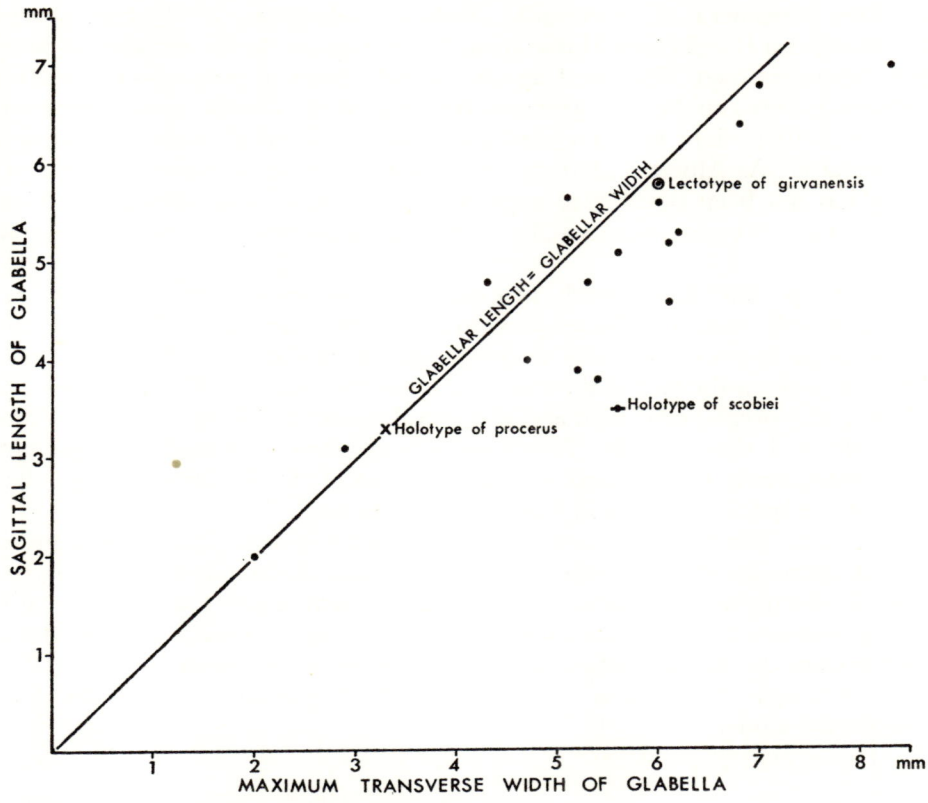

TEXT-FIG. 8. *Paraproetus girvanensis* (Nicholson & Etheridge, 1879).
Variation of glabellar length to width. The 'species' *procerus* and *scobiei* can be seen to fall into a continuously variable series.

rings; pleural areas with 6 or 7 pairs of ribs with flat-topped profile; pygidial border may be developed; sculpture of granules, discontinuous raised ridges, or smooth.

Remarks. The above combination of characters serves to distinguish the Warburgellinae from the Tropidocoryphinae. The origins of the Warburgellinae appear to lie in *Astroproetus* or a similar tropidocoryphine, as the pygidial pleural rib-structure and other characters of this genus (see above) show some similarities to warburgellines.

The Warburgellinae are known from the Silurian and earliest Devonian, reaching their greatest diversity in the later Silurian. *Koneprusites* Přibyl, 1964 (type species *K. moestus* (Barrande, 1852)), which occurs in the Middle Devonian of the Rheinisches Schiefergebirge, Germany, in Morocco and in the Prague district of Czechoslovakia, has pygidial characters (narrow axis, flat-topped type of pleural ribs) which are similar to those of warburgellines, although the glabella lacks the typical deep 1p furrows. It is possible that *Koneprusites* is a late warburgelline, but information on the rostral plate is required before this can be confirmed.

Genus **WARBURGELLA** Reed, 1931
(= *Podolites* Balashova, 1968. ?*Waigatchella* Maksimova, 1970)

Type species. Originally designated by Reed 1931, p. 14; *Asaphus Stokesii* Murchison, 1839, p. 656; from the Wenlock Limestone (Wenlock Series), of Dudley, Worcestershire.

Diagnosis. Cephalon with or without tropidium or tropidial ridges; preglabellar field short (sag.), commonly with transverse preglabellar ridge; lateral occipital lobes typically well developed; thorax of 8–10 segments; pygidium with or without well developed border; sculpture of granules or continuous ridges, or exoskeleton smooth.

Remarks. Two subgenera of *Warburgella*, *Podolites* Balashova, 1968 (type species *Cyphaspis rugulosa* Alth, 1874) and *Waigatchella* Maksimova, 1970 (type species *W. yakovlevi* Maksimova, 1970) have recently been proposed, diagnosed mostly on differences of proportion, sculpture, etc. The principal differences between the type species of *Warburgella* and *Podolites* are the number of thoracic segments (ten in former, eight in latter) and the lack of a pygidial border in the former and its presence in the latter. As Alberti (1970, p. 78) indicated, these characters are not of subgeneric importance. It has not been possible to assess fully the genus *Waigatchella*, but from published figures (Maksimova 1970, pl. 1, figs. 10–28), it is doubtful whether it should be separated from *Warburgella*.

Chlupáč (1971, p. 163) has recently proposed the subgenus *Prantlia* (*Tetinia*), with *Prantlia minuta* Přibyl & Vaněk, from the Lower Devonian (Lochkovian) of the Prague district, Czechoslovakia, as the type species. The only major morphological feature distinguishing this species from *Warburgella* species is the lack of the tropidium, but the presence of the transverse preglabellar ridge, the short (sag.) preglabellar field and the overall structure of the pygidium all distinguish it from *Prantlia* species. I thus consider *Tetinia* to be more closely related to *Warburgella* than to *Prantlia*, and consequently here regard it as a subgenus of the former instead of the latter, and would refer to it the species *minuta* and *ludlowensis* (see below).

The generic name *Warburgella* is commonly found in Ordovician faunal lists, but examination of species of this age assigned to it has shown them to belong to the morphologically similar *Cyphoproetus*. Ways by which the two genera can be distinguished can be found in the table on p. 28. From available evidence *Warburgella* first appears in Llandovery times, the earliest known representatives being from the Idwian Stage. The last representatives, subspecies of *W. rugulosa* (Alth, 1874) achieve very wide distribution at the base of the Devonian, at about the horizon of the *Monograptus uniformis* Zone.

SUBGENUS	Species	Transverse preglabellar ridge	Tropidium, tropidial ridges	Number of thoracic segments	Number of pygidial axial rings
WARBURGELLA	stokesii	present	tropidium present	10	8–9
WARBURGELLA	scutterdinensis	absent	tropidial ridges present	?	?6
WARBURGELLA	sp.1	?	?	10	7
TETINIA	ludlowensis	present	absent	8	9–10

TABLE 6. Summary of diagnostic characters of *Warburgella* (*Warburgella*) and *Warburgella* (*Tetinia*) species described herein.

Subgenus **WARBURGELLA** Reed, 1931

Type species. As for genus.

Diagnosis. Tropidium or tropidial ridges always present; transverse preglabellar ridge commonly present; thorax of 8–10 segments.

1. **Warburgella** (**Warburgella**) **stokesii** (Murchison, 1839)   Pl. 13, figs. 5–13; Pl. 14, fig. 2

	1839	*Asaphus Stokesii* Murchison, p. 656, pl. 14, fig. 6.
	?1851	*Proetus Stokesi* (Murchison); Angelin, p. 21, pl. 17, figs. 4, 4a–c.
	?1854	*Phaëthonides Stokesi* (Murchison); Angelin, p. 22.
	1854	*Proetus Stokesii* (Murchison); Murchison, pl. 17, fig. 7.
	1854	*Forbesia Stokesii* (Murchison); McCoy, p. 174.
	1854	*Proetus Stokesii* (Murchison); Morris, p. 114.
	1873	*Proetus Stokesii* (Murchison); Salter, p. 134.
	1877	*Proetus Stokesii* (Murchison); Woodward, p. 56.
	?1885	*Phaetonides Stokesi* (Murchison); Lindström, p. 75.
non	1904	*Proetus stokesi* (Murchison); Reed, p. 79, pl. 11, figs. 10, 11 [=*Cyphoproetus externus* Reed].
non	1914	*Proetus stokesi* (Murchison); Jones *in* Strahan *et al.*, pp. 108, 246 [=*Proetus haverfordensis* sp. nov.].
	?1916	*Proetus?*; Thomas *in* Cantrill *et al.*, pp. 76, 174.
	1916	*Proetus stokesi* (Murchison), var. nov. *bellula*, Reed, p. 165, pl. 8, figs. 5–10, *non* fig. 11 [=phacopid].
	1925	'*Phaetonides*' *Stokesi* (Murchison); Warburg, p. 184.
	1927	*Proetus Stokesii* (Murchison); Kegel, p. 637.
	1931	*Proetus* (*Warburgella*) *stokesi* (Murchison); Reed, p. 14.
	1935	*Proetus stokesi* (Murchison); Reed, p. 42.
	1938	*Warburgella stokesi* (Murchison); Whittard, p. 95, pl. 3, fig. 1 *non* fig. 2 [=*Warburgella* sp. 1], *nec* fig. 3 [=*Cyphoproetus binodosus* (Whittard)].
	1961	*Proetus* sp.; Mitchell, Pocock & Taylor, p. 30 (*partim*).
	1963	*Warburgella* cf. *baltica* Alberti, p. 155, pl. 15, figs. 10, 11, pl. 16, fig. 13.
	1967	*Warburgella stokesi* (Murchison); Ormiston, p. 62.
	1969	*Warburgella stokesii* (Murchison); Alberti, p. 354.
	1969	*Warburgella baltica* Alberti; Alberti, pp. 354, 456 (*partim*), pl. 33, fig. 15.
	1970	*Warburgella baltica* Alberti; Alberti, p. 78 (*partim*).
	1970	*Warburgella stokesii* (Murchison); Alberti, p. 78, pl. 15, fig. 20.
	1972	*Warburgella* cf. *stokesii* (Murchison); Bassett, p. 31.

Neotype. Proposed by Whittard 1938, p. 95; BU 176 (*ex* Ketley Collection no. 335), Pl. 13, figs. 5a–c; an almost complete specimen, lacking the right-hand free cheek, figured Whittard 1938, pl. 3, fig. 1; from Wenlock Limestone (Wenlock Series), Dudley, Worcestershire.

Material, horizons and localities. In the British Isles this species has only been recorded from the topmost Wenlock Shale and from the Wenlock Limestone. From the Wenlock Shale: cranidia, free cheeks and pygidia, GSM Zi2807–8, Zi2811, Zi2819, Zi2820, Zi2822, Zi2829 from Harley Hill, 1 mile SE of Harley, Shropshire. From Wenlock Limestone: numerous complete or nearly complete specimens in museum collections labelled 'Wenlock Limestone, Dudley' (e.g. SM A28255–62, BU 1838, 1839, NMW 93.219), preserved either in limestone or grey mudstone. Apart from these, there are numerous detached exoskeletal parts, including: NMW 71.6G.232–238 from Nodular Beds, large bedding plane exposures 230–270 yd SW of 'Caves' public house (SO 9350 9210) and LRU 53758, 450 yd SSW (SO 9345 9180) of 'Caves' public house, Wren's Nest Hill, Dudley, Worcestershire; NMW 71.6G.327, near top of nodular facies, old roadside quarry in Harton Hollow Wood, *c.* $\frac{3}{4}$ mile S of Harton, Wenlock Edge (SO 4085 8762); GSM Dr907, 908, 913, 932, from old quarry 130–150 yd S of Fetterlocks Farm, 2360 yd ESE of Shelsley Beauchamp church, Worcestershire (SO 752 632); NMW 72.18G.66, old quarry 450 yd WSW of Gold Hill Farm, near Eastnor, Herefordshire (SO 7325 3631); NMW 71.6G.498–499 and LRU 53759–60, Clenchers Mill, $2\frac{1}{4}$ miles SE of Ledbury church, Herefordshire (SO 735 376); SM A16595–8, A16600–01, (syntypes of *W. stokesii* var. *bellula*), stream section near quarry 300 yd NE of Greenpool Farm, 1500 yd N of church at Common Coed-y-paen, Usk, Monmouthshire (ST 3343 9991) [now flooded by Llandegfedd Reservoir]; GSM DEX2927, old quarry opposite Ty-newydd Farm, 1400 yd at 127° from Llanarthney church, Carmarthenshire (SN 5442 1951); a *Warburgella* species, possibly *W. stokesii* (GSM HT1262) occurs in the late Llandovery or Wenlock of Deadman's Bay, Pembrokeshire. *W.* (*W.*) *stokesii* is also recorded from erratics of Wenlock age from Hiddensee Island, Germany (see below).

Diagnosis. Prominent transverse preglabellar ridge; tropidium present; lateral occipital lobes distinct; genal spine extends backwards as far as ninth thoracic segment; thorax of 10 segments;

pygidium without border, axis with 8 or 9 rings, pleural areas with 5 or 6 pairs of ribs; sculpture of fine discontinuous raised ridges.

Description. Cephalon moderately vaulted with broad, weakly convex border defined by shallow but distinct anterior and lateral border furrows. Cranidium with sagittal length somewhat greater than palpebral width. Glabella trapezoidal, moderately convex in longitudinal and transverse sections, bluntly rounded anteriorly, widest across δ–δ, and distinctly constricted opposite anterior end of eye. Three pairs of lateral glabellar furrows; 1p: deep, prominent, abaxial end shallow, opposite anterior part of palpebral lobe, from here deepening rapidly backwards but shallowing again before reaching occipital furrow. Directed obliquely backwards at about 45°, partially isolating ovate 1p lobe which has independent convexity from remainder of glabella. 2p and 3p both faint, not incised, only seen on well preserved specimens (e.g. Pl. 13, fig. 13b). 2p: opposite γ, directed weakly backwards; 3p: short distance in front of 2p, running nearly transversely.

Occipital furrow rather shallow, running more or less straight (trans.) Occipital ring approximately same width (sag.) as anterior border, maintaining more or less constant width laterally, and at extreme lateral ends are small, ovate lateral occipital lobes. Small median tubercle present. Occipital ring as wide (sag.) as preglabellar field. Preglabellar field traversed by tropidium a short distance in front of glabella, and between tropidium and anterior border furrow is a prominent elongated transverse preglabellar ridge. The latter is damaged and difficult to see on the neotype. Section β–γ of anterior branch of facial suture diverges abaxially forwards at 30°–40° from γ, α–β–γ describing sigmoidal curve. Posterior branches with ϵ and ξ as a single broad curve.

Palpebral lobe backwardly placed, close to axial furrow, semi-elliptical in outline, about $\frac{1}{2}$ or a little over $\frac{1}{2}$ the sagittal length of glabella, and inclined at a low angle from axial furrow in sagittal view. Eye large, crescentic. No distinct eye socle. Field of free cheek traversed by tropidium, which runs parallel to margin and makes a distinct break in slope; adaxially from it field of free cheek gently declined, abaxially from it, it is steeply declined. Posterior border furrow deeper and narrower than lateral, posterior border narrower than lateral, inclined towards posterior. Genal spine broad-based, extending backwards as far as ninth thoracic segment. It has a distinct median groove which gradually shallows towards posterior, dividing it into inner and outer bands of approximately equal width (trans.).

Hypostome oblong, median body convex, rising to crest near anterior end. Border furrows narrow but distinct, anterior border narrow, upturned, lateral and posterior borders wider, with two or three prominent raised ridges running parallel to margin. Anterior wings badly preserved on single available specimen. Maculae, situated close to posterior end, are prominent.

Thorax of ten segments, axis tapering backwards so that last ring is about $\frac{2}{3}$ width of first (trans.). Each axial ring arched weakly forwards sagittally, in lateral profile annulus weakly convex, gently inclined towards posterior. Pleurae nearly transverse, curving gently downwards at fulcrum, which is closer to adaxial than to abaxial end for entire length of thorax. Articulating facet transversely elongated. Pleural furrow narrow, dividing pleura into anterior and posterior bands of approximately equal width (exsag.), dying out a short distance from abaxial end. Posterolateral corner of pleura rounded.

Pygidium without border, subparabolic in outline. Axis narrow, anteriorly about 28% of pygidial width, tapering gently backwards, strongly convex longitudinally, with eight or nine rings which are weakly convex in longitudinal profile and defined by shallow ring furrows. Pleural areas broad, gently convex, with five or six pairs of ribs which curve gently backwards, widening slightly abaxially. Pleural furrows narrow, incised, reaching close to margin. Interpleural furrows weak, more or less parallel with pleural until close to their abaxial ends where they become incised and more strongly backwardly turned, approaching as close to margin as pleural furrows.

Sculpture of fine discontinuous raised ridges and granules, most prominent on glabella, palpebral lobe and on pygidium.

Measurements.

Cranidia	A	A_1	A_2+A_3	A_4	K	$\delta-\delta$	
BU 176 (E)	5·2	3·0	1·4	0·8	3·1	4·5	NEOTYPE
SM A28260 (E)	—	5·2	1·5	—	4·9	6·0	
SM A28261 (E)	6·2	4·1	1·4	0·7	3·9	5·1	
SM A28255 (E)	5·9	3·7	1·6	0·6	3·7	(4·2)	
SM A28256 (E)	5·8	3·6	1·5	0·7	3·3	4·4	
SM A28257 (E)	5·5	3·4	1·2	0·9	3·3	4·2	
SM A28262 (E)	4·9	3·3	0·9	0·7	2·6	3·8	
SM A16595 (E)	—	3·0	—	0·6	2·8	—	
BU 1838 (E)	4·7	2·9	1·4	0·4	2·4	3·3	
BU 1839 (E)	4·3	2·4	1·2	0·7	2·4	3·0	
SM A16596 (E)	4·0	2·3	1·1	0·6	2·1	3·2	

Pygidia	Z	Y	W	X	
BU 176 (E)	4·0	3·0	6·1	1·8	NEOTYPE
SM A28256 (E)	3·9	3·0	6·8	2·0	
SM A28262 (E)	3·4	3·0	5·7	1·8	

Remarks. Alberti (1969, pp. 354–5; 1970, p. 78) has stated that the number of thoracic segments in the type species of *Warburgella* is nine, but in fact this species has ten. Whittard (1938, p. 96) noted that the cephalon of the neotype overlaps the anterior part of the thorax, so that only nine thoracic segments are visible (see Pl. 13, fig. 5a), but he suspected ten to be the full complement, and his suspicion is confirmed, following examination of numerous complete specimens (e.g. Pl. 13, figs. 6, 7), all of which show ten.

Reed (1916, p. 165) described *W. stokesii* var. *bellula* on the basis of specimens from the Usk Inlier, using as characters to distinguish it from *W. stokesii stokesii* "the relatively shorter and blunter glabella, the stronger development of the first and second glabellar furrows and the larger eyes". Such differences are, apparently, merely products of variation and preservation. A range of glabellar shapes can be seen in Pl. 13, figs. 6, 7, 11 and 13, and the 2p and 3p furrows are not seen where the preservation is poorer—e.g. Pl. 13, fig. 5, but are clearly seen in well-preserved specimens—e.g. Pl. 13, figs. 7, 11, 13. Likewise the transverse preglabellar ridge is difficult to see in some specimens—e.g. Pl. 13, fig. 5, which has been slightly crushed at the anterior end. For these reasons, *bellula* is not recognised as a subspecies of *W. (W.) stokesii*.

In the collections of the British Museum (Natural History) there is a single specimen of *Warburgella* (BM I4520) labelled as having originated from "Wenlock Shale, Wenlock Edge". This specimen (see Pl. 14, figs. 1a–c) differs from typical *W. stokesii* in its distinctly flattened glabella and narrower inner portion of the free cheek. These differences may be specific, or possibly due to extreme variation, but further material is required before definite pronouncement can be made, and here the specimen is referred to as *W. aff. stokesii*.

From Gotland, Lindström (1885, p. 75, pl. 16, fig. 13) described and figured *Phaëtonides* [=*Warburgella*] *rugulosus* Lindström (refigured here Pl. 14, figs. 3a–b). Because *W. rugulosa* (Lindström, 1885) is a junior homonym of *W. rugulosa* (Alth, 1874), Alberti (1963, p. 150) proposed the new name *baltica* for the former, and he figured (*op. cit.*, pl. 15, fig. 16) specimens from Wenlock erratics from Hiddensee Island, N of Stralsund, E Germany, as *W. cf. baltica*. Alberti (1969, pl. 33, fig. 15) later designated one of the Hiddensee cranidia as the 'holotype' of *baltica*. Such a designation is, however, nomenclatorially incorrect, as Lindström's monotypic cranidium (RM Ar28751) of *rugulosus* is the holotype of *baltica*, and also as specimens doubtfully attributed to a species cannot be designated as a lectotype of that species. Both *W. (W.) stokesii* and Alberti's Hiddensee specimens differ from the type of *baltica* in possessing the transverse preglabellar ridge as well as in details of the lateral glabellar furrows and sculpture (cf. Pl. 13, fig. 13; Pl. 14, figs. 3a, b herein and Alberti 1963, pl. 15, figs. 10, 11a), and the latter species is clearly specifically distinct from the others. There is little to distinguish the Hiddensee specimens from *stokesii*, and they are here assigned to that species. The differences which Alberti (1970, p. 79) noted between the Hiddensee specimens (upon which his concept of *baltica* was based) and *stokesii* are ones only of degree—e.g. the lateral profile of the glabella and width of the preglabellar field, which in my opinion can be accounted for by intra-specific variation.

2. **Warburgella (Warburgella) scutterdinensis** sp. nov.   Pl. 14, figs. 4–7; Text-fig. 9

Name. From the type locality, near Woolhope, Herefordshire.

Type specimens. Holotype, NMW 72.18G.70, cranidium; Pl. 14, figs. 6a–c; from Woolhope Limestone (Wenlock Series), old quarry at Scutterdine, 1000 yd SE of church at Mordiford, Herefordshire (SO 5775 3686); paratypes, NMW 72.18G.71, free cheek (Pl. 14, fig. 5) and NMW 72.18G.67, pygidium (Pl. 14, fig. 7) from type locality; NMW 29.62G.27, free cheek (Pl. 14, fig. 4) labelled "Woolhope Limestone, Woolhope".

Material, horizon and locality. Besides the type specimens, NMW 72.18G.68, pygidium, and NMW 72.18G.69, free cheek, both from the type locality.

Diagnosis. Glabella weakly inflated; no transverse preglabellar ridge; tropidial ridges developed; lateral occipital lobes poorly developed; palpebral lobe elevated to height of sagittal region of glabella; median tubercle close to posterior edge of occipital ring; pygidium without border, and with ?6 axial rings and ?5 pairs of pleural ribs.

Description. Cephalon rather weakly inflated; broad, rather weakly convex border defined by shallow anterior and lateral border furrows. Glabella as long (sag.) as wide (trans.), weakly inflated, tapering gently forwards and bluntly rounded in anterior. Shallow, but distinct 1p furrow runs obliquely backwards at 30° from where it meets axial furrow about $\frac{1}{2}$ way up side of glabella, shallowing and becoming indistinct before reaching occipital furrow. Resultant 1p lobe triangular and very weakly inflated. 2p inconspicuous, transverse, running into axial furrow a short distance in front of 1p. 3p similar to 2p, meeting axial furrow a short distance in front of it.

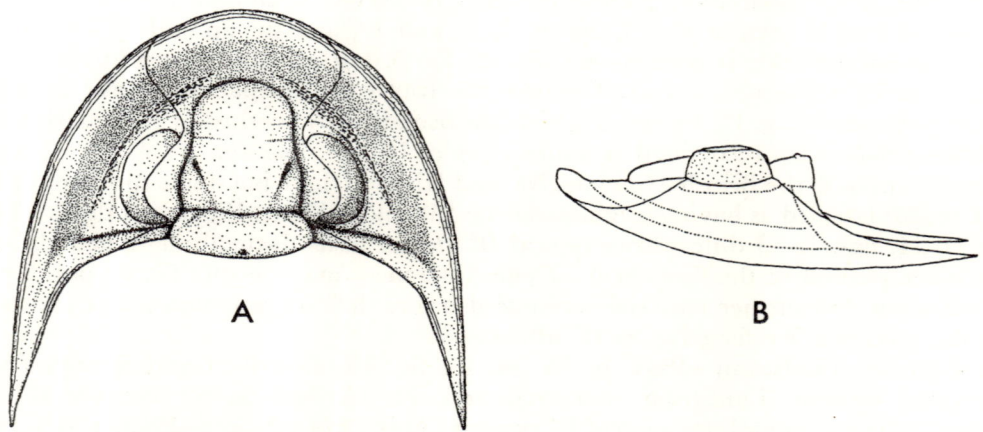

TEXT-FIG. 9. Reconstruction of the cephalon of *Warburgella (Warburgella) scutterdinensis* sp. nov. A. Dorsal view, B. Lateral view, showing depressed glabella. Based on Pl. 14, figs. 4–6. × 10 approx.

Occipital furrow with vertical anterior slope, arched weakly backwards between adaxial ends of 1p lobes. Lateral ends directed weakly backwards. Occipital ring about $\frac{1}{2}$ length (sag.) of preglabellar area and a little wider (trans.) than glabella. Lateral occipital lobes only weakly developed. Prominent median tubercle situated close to posterior edge of occipital ring.

Preglabellar field concave in profile and about 57% length (sag.) of glabella. Anterior border about $\frac{2}{3}$ length (sag.) of preglabellar field. Section β–γ of anterior branch of facial suture diverges abaxially forwards at 38° from γ, which is close to axial furrow, not far from anterior end of 1p furrow. Palpebral lobe large, about 64% sagittal length of glabella, with abaxial part elevated to height of sagittal region. Eye not preserved on any specimen. Eye socle with non-incised lower margin which diverges weakly from upper at either end. Posterior branch of facial suture with ϵ and ξ as a single, wide angle, close to axial furrow.

Tropidial ridges on a strong fold which traverses field of free cheek and lateral part of pre-glabellar field, dying out in region in front of glabella. Inner part of free cheek very narrow, outer part concave, trough-like, merging insensibly with lateral border furrow.

Posterior border furrow narrow and distinct. Genal spine broad-based, with median groove slightly off-set abaxially from lateral border furrow.

Hypostome and thorax unknown.

Pygidium subparabolic, without border. Axis with ?six rings, defined by shallow ring furrows. Pleural areas broad, with ?five pairs of ribs. Pleural furrows narrow, distinct, maintaining constant depth along their length and reaching close to margin. Interpleural furrows weak, apparent towards their abaxial ends, where they curve backwards more strongly than the pleural furrows. Anterior and posterior pleural bands of more or less equal width (exsag.) and convexity.

Most of exoskeleton smooth, but locally developed granules occur on some parts, e.g. on posterior part of glabella.

Measurements.

Cranidium	A	A_1	A_2+A_3	A_4	K	$\delta-\delta$	
NMW 72.18G.70 (E)	3·5	1·8	1·0	0·7	1·7	—	HOLOTYPE

Pygidium	Z	Y	W	X		
NMW 72.18G.67 (E)	—	—	2·9	(0·6)	PARATYPE	

Remarks. W. (W.) *scutterdinensis*, unlike W. (W.) *stokesii*, lacks the transverse preglabellar ridge and has tropidial ridges rather than a tropidium. The weak inflation of the glabella of the former recalls W. (W.) aff. *stokesii* (Pl. 14, figs. 1a–c), but W. (W.) aff. *stokesii* (like W. (W.) *stokesii*) differs from W. (W.) *scutterdinensis* in having a transverse preglabellar ridge and a tropidium rather than tropidial ridges.

The lack of a transverse preglabellar ridge in W. (W.) *scutterdinensis* is a feature in common with W. (W.) *baltica* (Pl. 14, figs. 3a–b), although other features of the latter—the tropidium and glabellar inflation, are like W. (W.) *stokesii*. W. (W.) sp. 1 (see below), like W. (W.) *scutterdinensis*, has tropidial ridges.

3. **Warburgella (Warburgella) sp. 1**   Pl. 14, figs. 8–12

1938   *Warburgella stokesi* (Murchison) *partim*; Whittard, p. 95, pl. 3, figs. 2, 2a, *non* figs. 1, 3.

Material, horizons and localities. This species has been recorded only from the Llandovery Series. From the Idwian Stage: Venusbank Formation—OUM C11328, cranidium, OUM C11318, C11321, C11325, free cheeks and OUM C11311, pygidium, from Josey's Wood, *c.* 2 miles SSW of Minsterley, Shropshire (SG 3653 0221) [ZCM loc. 73]; OUM C9312, C9315–16, free cheeks and OUM C9308, pygidium from Hope Quarry, *c.* 2¼ miles SW of Minsterley (SJ 3550 0207) [ZCM loc. 70]; Bog Quartzite—BM It8264, It8266, cranidia, BM It8268, free cheek and BM It8267, pygidium from Round Hill, Bank Outlier, Shropshire (SO 3500 9924). From Fronian–Telychian Stage, Pentamerus Beds: GSM 55470–71, complete specimen (counterpart internal and external moulds), from ⅙ mile N of Spout Lane, near Little Wenlock, Shropshire (SJ 634 072); OUM C15380, pygidium, from Morrellswood, 1¼ miles WSW of Little Wenlock (SJ 6284 0637); OUM C14621, pygidium, from 400 yd ESE of Merrishaw, 1500 yd SW of church at Harley, Shropshire (SJ 5852 0065) [ZCM loc. 54].

Description. None of the above material is very well preserved, and most of the information is derived from GSM 55470–71 the complete specimen figured by Whittard (1938, pl. 3, figs. 2, 2a), refigured herein Pl. 14, fig. 10. Glabella trapezoidal, similar to that of W. *stokesii*. Occipital ring with small lateral occipital lobe, distinctly smaller than that of W. *stokesii*. Tropidium a fold rather than a clearly defined ridge as in W. *stokesii*. Genal spine without deep median groove. Thorax, like *W. stokesii*, of ten segments. Pygidium with seven axial rings, five pairs of pleural ribs. Sculpture of fine raised striations, clearly seen on free cheeks OUM C9312 (Pl. 14, fig. 8) and C9315.

Measurements.

Cranidia	A	A_1	A_2+A_3	A_4	W	$\delta-\delta$
GSM 55470 (I)	—	3·2	—	0·7	(2·8)	—
OUM C11328a (I)	(5·1)	3·2	1·1	(0·8)	(3·0)	—

Pygidia	Z	Y	W	X
GSM 55470 (I)	2·7	2·2	(5·9)	1·3
OUM C15380a (E/I)	7·1	5·6	—	2·8
OUM C14621 (E)	1·8	1·4	3·5	0·8

Remarks. The salient features of this Llandovery species are noted in the above description, but a fuller description must await better material. As far as can be seen, one species is represented which differs from *W. (W.) stokesii* in having tropidial ridges, smaller lateral occipital lobes and in having fine striations on the cheeks. The tropidial ridges and small lateral occipital lobes are characters shared with *W. (W.) scutterdinensis*. *W. (W.)* sp. 1 is the earliest, and the only Llandovery species, of *Warburgella* known to the author.

Subgenus **TETINIA** Chlupáč, 1971

Types species. Originally designated by Chlupáč 1971, p. 163; *Prantlia minuta* Přibyl & Vaněk, 1962, p. 26; from Lochkov Series (Lower Devonian), Stydlé vody, Loděnice, Prague district, Czechoslovakia.

Diagnosis. Tropidium absent; preglabellar field with preglabellar ridge; thorax of 8 segments.

Warburgella (Tetinia) ludlowensis (Alberti, 1967) Pl. 14, figs. 13?, 14–18; Pl. 15, figs. 1, 2; Text-fig. 10

1967 *Proetus* (sg.?) *ludlowensis* Alberti, p. 483, pl. 1, fig. 2.
1969 *Proetus*? (sg.?) *ludlowensis* Alberti; Alberti, p. 368, pl. 33, figs. 16a, b.
1969 *Warburgella*? sp. (aff. *Warburgella rugulosa* (Alth)); Alberti, pl. 46, fig. 18.
1970 *Warburgella*? *ludlowensis* (Alberti); Alberti, p. 78.

Type specimens. Holotype, SMF 23363, an incomplete cranidium, figured Alberti 1967, pl. 1, fig. 2, and 1969, pl. 33, figs. 16a, b, a cast of which is refigured herein Pl. 14, fig. 14; from Lower Leintwardine Beds (Ludlow Series), roadside exposure ½ mile ENE of Mary Knoll House and 1½ miles WSW of Ludlow church, Shropshire (SO 4887 7389). Paratype, SMF 27145, fragmentary cranidium from Lower Leintwardine Beds, Church Hill, Leintwardine. Besides this specimen, Alberti designated SMF 23364, a cranidium, and SMF 27144 and SMF 27146, pygidia, as paratypes. However, from examination, these specimens are better assigned to *Proetus (Lacunoporaspis) obconicus*.

Material, horizons and localities. This species occurs in small numbers in the Leintwardine Beds in the Welsh Borderland. The following are recorded from the Lower Leintwardine Beds: NMW 71.21G.1, complete internal mould from quarry debris on S side of forestry track on SE side of Mary Knoll Valley, 2 miles SW of Ludlow (SO 489 725); BM It8818, cranidium, from trackside exposure in lane from Whitbach Farm to Hungerford, Wenlock Edge district (SO 5340 8984); BM It8849, pygidium, from quarry at SE end of Bache Plantation, Siefton Batch, Wenlock Edge district (SO 4772 8477); NMW 72.23G.1, cranidium from higher part of Cwm Jenkin, *c*. 3 miles WNW of Knighton, Radnorshire (SO 235 730). From the Upper Leintwardine Beds: NMW 71.6G.179, cranidium from 'Goggin' lane section, 1 mile SE of Elton, Herefordshire (SO 472 702); LCM 321'1970/1, cranidium from Lawnwell Dingle, near Leintwardine (SO 4167 7678); BM It8817, It8819, cranidia, from quarry on Diddlebury–Middlehope road, 220 yd NE of Fernhall Mill, Wenlock Edge district (SO 5006 8666); cranidium from field exposure near stream, 1700 yd NNW of church at Llangybi, Monmouthshire (ST 3679 9812). Besides these specimens, there are several nearly complete specimens, e.g. BM 39407, GSM 36747, in old museum collections, labelled from 'Leintwardine' or 'Church Hill'. They probably originated from Leintwardine Beds.

Diagnosis. Transverse preglabellar ridge weak; lateral occipital lobes distinct, well developed; pygidial border poorly developed; pygidial axis with 9–10 rings; pleural areas with 6 pairs of ribs; sculpture of fine granules.

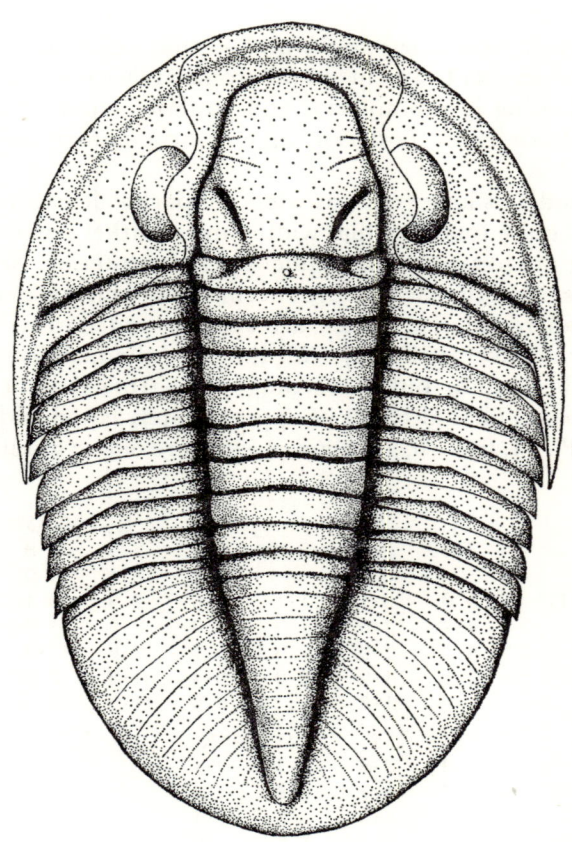

TEXT-FIG. 10. Reconstruction of *Warburgella* (*Tetinia*) *ludlowensis* (Alberti, 1967), based on Pl. 14, figs. 16, 17, and Pl. 15, fig. 1. The pygidium of this reconstruction is based on a variant (Pl. 15, fig. 1) with a rather elongated pygidium. Other specimens (e.g. Pl. 14, fig. 15) have a proportionately shorter pygidium. × 6 approx.

Description. Cephalon parabolic, with rather narrow border defined by shallow lateral and anterior border furrows. Glabella trapezoidal, weakly constricted laterally with a bluntly rounded frontal lobe. Three pairs of lateral glabellar furrows; 1p: deep, adaxially convex, running into axial and occipital furrows, shallowing markedly at either end. 2p: weak, running slightly backwards, situated at point of greatest constriction of glabella, a short distance behind γ. 3p not far in front of 2p, of similar nature and length, directed slightly forwards. Frontal lobe about $\frac{1}{3}$ sagittal length of glabella. Glabella rather weakly inflated, with basal lobes having independent convexity from remainder.

Occipital furrow deep, curving forwards weakly laterally, where it is deepest. Occipital ring a little less than $\frac{1}{4}$ total sagittal length of glabella, and is slightly wider (trans.). Small median tubercle present. Lateral occipital lobes prominent, incompletely isolated by furrows deepening and running obliquely forwards adaxially. Preglabellar field short (sag.) with distinct preglabellar ridge, defined by weak furrows in anterior and posterior.

Section β–γ of anterior branch of facial suture diverges abaxially forwards at 19°–24° from γ, which is close to axial furrow. Palpebral lobe subcrescentic, a little under $\frac{1}{2}$ sagittal length of glabella. Eye crescentic, with narrow, indistinct eye socle. Posterior branches of facial sutures with ϵ and ξ as single angle.

Field of free cheek broad, weakly convex. Genal spine rather narrow, extending as far back as fifth thoracic segment. Posterior border furrow narrow, deep and distinct, abruptly truncated at base of genal spine.

Thorax of eight segments. Axis tapers backwards so that eighth ring is $\frac{3}{4}$ width of first. Axial rings arched gently forwards sagittally and in lateral profile are weakly convex. Pleura with narrow, distinct pleural furrow, dividing it into anterior and posterior bands of more or less equal width (exsag.) and truncated abaxially by posterior edge of articulating facet. Posterolateral corner angular.

Pygidium between $\frac{1}{2}$ and $\frac{2}{3}$ as long (sag.) as wide, subparabolic, with a poorly defined border. Anteriorly axis occupies about $\frac{1}{3}$ total pygidial width, and is rather narrow, with nine to ten rings, the ring furrows becoming progressively less distinct towards posterior. No distinct postaxial ridge. Pleural areas with six pairs of gently backwardly curving ribs, which widen slightly adaxially. First pleural furrow deeper than remainder, which are shallow and narrow and of similar depth to interpleural furrows, which converge slightly with the pleural adaxially. First pair of pleural and interpleural furrows reach margin, but succeeding pairs truncated at inner edge of pygidial border. Anterior and posterior pleural bands of approximately equal width (exsag.).

Some specimens (e.g. Pl. 14, figs. 16, 17) show a sculpture of fine granules.

Measurements.

Cranidia	A	A_1	A_2+A_3	A_4	K	$\delta-\delta$	
SMF 23363 (E)	(4·5)	2·9	(0·9)	0·7	2·9	3·8	HOLOTYPE
GSM 36747 (E)	—	3·8	—	0·6	3·4	(4·0)	
NMW 71.21G.1 (I)	5·0	3·1	1·1	0·8	2·9	(3·4)	
BM It8819 (E)	4·7	2·9	1·0	0·8	2·9	4·0	
BM It8818 (E)	4·5	2·9	0·9	0·7	(2·9)	(3·8)	
BM It8817 (E)	3·1	1·9	0·8	0·4	1·8	(2·3)	
BM 39407 (I)	—	1·5	—	0·3	1·5	2·0	

Pygidia	Z	Y	W	X
GSM 36747 (E)	4·8	3·5	7·3	2·1
NMW 71.21G.1 (I)	4·3	3·7	6·3	1·3
BM It8849 (E)	3·0	2·3	5·3	1·4
BM 39407 (I)	1·6	1·3	3·2	0·9

Remarks. There is a small incomplete cranidium, LCM 321′1970/17, from higher Lower Leintwardine channel-fill deposits from Marlow Lane, near Leintwardine (SO 4032 7678) (see Pl. 14, fig. 13) which has a sagittally widened anterior border and incised 2p and 3p furrows (1p not preserved); this specimen might be an early growth stage of *W. (T.) ludlowensis*.

Alberti (1969, pl. 46, fig. 18) figured a pygidium from the type locality as *Warburgella* aff. *W. rugulosa* (Alth), but comparison of this specimen with complete examples of *W. (T.) ludlowensis* indicates that it belongs to the latter.

W. (T.) ludlowensis is similar to *W. (T.) minuta* (see Chlupáč 1971, pl. 20, figs. 1–9; pl. 24, fig. 9), but differs from the latter in its better developed lateral occipital lobes, proportionately wider glabella, narrow cephalic border and less well developed pygidial border.

Genus **PRANTLIA** Přibyl, 1946

Type species. Originally designated by Přibyl 1946, p. 90; *Proetus longulus* Hawle & Corda, 1847, p. 76; from Kopanina Beds (Ludlow Series), Dlouhá Hora, Prague district, Czecholsovakia.

Diagnosis. Cephalon without tropidium; preglabellar field long (sag.), without transverse preglabellar ridge, lateral occipital lobes well to poorly developed; thorax of 9–10 segments; pygidial axis with 7–10 rings, pleural areas with 5–6 pairs of ribs; sculpture of fine striations, or exoskeleton smooth.

Remarks. *Prantlia* is an uncommon genus, with three species known to the author—*P. longula*

(Hawle & Corda) from the Ludlow of Czechoslovakia, *P. longifrons* (Lindström) from the Ludlow of Gotland and *P. grindrodi* sp. nov. from the Wenlock of the Welsh Borderland.

Prantlia grindrodi sp. nov.   Pl. 15, figs. 3–5

Name. The old manuscript name *grindrodi* is revived; in honour of R. B. Grindrod, who made extensive collections of Silurian fossils from the Malvern area.

1878   *Proetus grindrodi* Edgell MS; Huxley, Newton & Etheridge, p. 84 [*nom. nud.*].

Diagnosis. Lateral occipital lobes poorly developed; thorax of 10 segments; pygidium without border, axis with 7 rings, pleural areas with 5 pairs of ribs, interpleural furrows deepening abaxially; cephalic and pygidial doublures very broad.

Type specimens. Holotype, GSM 3303, complete specimen; Pl. 15, fig. 5; from Wenlock Shale (Wenlock Series), Malvern. Paratype, OUM C799, complete specimen; Pl. 15, fig. 3; from Wenlock Series, locality unknown.

Material, horizons and localities. Several complete or nearly complete specimens: from the Wenlock Shale—OUM C776, C777, C786, C796–98, C801 from "Malvern Tunnel", GSM 36153–54 from "the Wych, Malvern", BM 59013–14 from "Malvern"; from the Woolhope Shale —BU 1840, from "Malvern"; from the Wenlock Limestone—BM 42630 from "Malvern" (there are four specimens included under this number; one is *P. grindrodi*, the others are *Proetus* (*Proetus*) *concinnus*.

Description. Cephalon of parabolic outline, anterior and lateral border furrows broad and shallow, border rather narrow. Glabella trapezoidal, weakly constricted laterally and bluntly rounded anteriorly. 1p deep, running backwards at about 30°, deepest medially and shallowing at either end, with anterior end situated a short distance behind anterior end of palpebral lobe. Resultant partially isolated 1p lobe has independent convexity from remainder of glabella, is ovate and elongated exsagittally. 2p short, incised, transverse, opposite γ. 3p shorter than 2p, a short distance in front of it and transverse.

Occipital furrow shallow and almost transverse, narrow medially, widening laterally. Occipital ring about as wide (trans.) as glabella, maintaining constant width (sag. and exsag.). Small median tubercle, situated posteriorly. Small, ill-defined triangulate lateral occipital lobes present. Preglabellar area about $\frac{3}{4}$ length (sag.) of glabella. Preglabellar field gently sigmoidal. Not far in front of glabella is a weak, nearly transverse depression, which continues on to free cheek, running parallel with margin.

Section β–γ of anterior branch of facial suture diverges abaxially forwards at 35°–40° from γ, which is close to axial furrow near anterior end of glabella. Palpebral lobe large, crescentic, close to glabella, backwardly placed, and steeply inclined from axial furrow. Eye crescentic, about $\frac{2}{3}$ sagittal length of glabella. Eye socle inconspicuous, its non-incised lower margin running more or less parallel with upper. Posterior branches with ϵ and ξ a short distance apart, intervening stretch parallel with axial furrow. Inner part of field of free cheek weakly convex. Part beyond depression gently concave. Posterior border furrow deep, narrow and truncated at base of genal spine. Genal spine broad-based, blade-like, extending backwards as far as ninth or tenth thoracic segment, with a broad, shallow, median groove.

Cephalic doublure very broad, with distinct terrace lines parallel with margin. Rostral plate large, connective sutures diverging backwards and adaxially convex.

Thorax of ten segments. First six axial rings maintain more or less constant width (trans.), but behind sixth, thoracic axis tapers quite rapidly backwards. Articulating furrow arched gently forwards. Articulating half ring weakly convex and about $\frac{2}{3}$ width (sag.) of annulus. Pleura adaxially nearly horizontal, but beyond fulcrum is fairly strongly declined. Pleural furrow narrow and distinct, dividing pleura into a narrower anterior band and a wide posterior band, and truncated abaxially by posterior edge of articulating facet. Posterolateral corner of pleura angular.

Pygidium subparabolic, without border, about twice as wide (trans.) as long (sag.). Anteriorly axis about $\frac{1}{4}$ total pygidial width (trans.), tapering backwards to a blunt point. Weak postaxial ridge. Seven axial rings, defined by shallow ring furrows, which curve gently backwards sagittally. Pleural areas with five pairs of ribs, with gently sigmoidal pleural and interplerual furrows. Pleural furrows maintain constant width along their length, and reach close to margin. Interpleural furrows more or less parallel with pleural, deepening slightly abaxially, where they are almost as deep as pleural furrows. Pygidial doublure broad, weakly convex ventrally, with distinct, parallel terrace lines.

Exoskeleton smooth.

Measurements.

Cranidia	A	A_1	A_2+A_3	A_4	K	δ–δ	
GSM 3303 (E)	5·6	2·8	2·0	0·8	2·7	(3·0)	HOLOTYPE
OUM C798 (E)	6·9	3·7	2·4	0·8	(3·3)	(4·5)	
OUM C776 (E)	6·4	2·9	2·6	0·9	3·1	(3·5)	
BM 59013 (E)	6·2	3·0	2·3	0·9	3·3	(4·3)	
OUM C799 (E/I)	5·4	2·7	2·2	0·5	(2·7)	(3·2)	PARATYPE

Pygidia	Z	Y	W	X	
GSM 3308 (E/I)	3·1	2·0	5·7	1·2	HOLOTYPE
OUM C798 (E)	4·2	3·9	7·9	2·0	
BM 59013 (E)	3·6	3·2	7·9	2·0	
OUM C799 (E/I)	3·0	2·6	5·9	1·4	PARATYPE

Remarks. Prantlia grindrodi differs from the type species, *P. longula*, in its less distinct lateral occipital lobes, more transverse frontal glabellar margin, in its extra thoracic segment and in lacking the pygidial border. The figure of *longula* in the *Treatise* (p. O396, fig. 301, 6) is, as Chlupáč (1971, p. 163) has pointed out, inaccurate and bears little resemblance to the specimen upon which it is based. The same is true of Pillet's (1969, pl. 4, fig. 12) illustration of the same species. Good photographs of the species have, however, recently been published by Chlupáč (1971, pl. 20, figs. 10–11) and by Horný & Bastl (1970, pl. 10, fig. 8) to which reference may be made.

Examination of a single, incomplete cephalon of *P. longifrons* suggests that the third species of *Prantlia*, *P. longifrons* (Lindström, 1885) from the Hemse Beds (Ludlow) of Gotland is closely comparable with *P. grindrodi* (cf. Pl. 15, figs. 3–5 and 6a–c). *P. longifrons* appears to lack the depression on the preglabellar field and free cheek, and has a sculpture of fine, raised striations disposed in a scaly arrangement; it is otherwise closely similar to *P. grindrodi*.

Subfamily Uncertain
Genus **RORRINGTONIA** Whittard, 1966

Type species. Originally designated by Whittard 1966, p. 292; *Rorringtonia flabelliformis* Whittard, 1966, p. 292; from the Rorrington Beds (Ordovician, Caradoc Series), Rorrington, Shropshire.

Diagnosis. Trapezoidal glabella with three pairs of incised furrows, 1p typically forked adaxially; large preglabellar field; posterior portion of fixed cheek triangular; short eye ridges may be present; occipital ring sagittally widened; thorax of 10 segments; pygidium with 10 axial rings and 9 pairs of pleural ribs, pleural and interpleural furrows curving progressively more strongly backwards.

Remarks. Whittard (1966, p. 292) regarded *Rorringtonia* as *incertae sedis*. He compared its cephalic characters with those of *Parabolinella* and its pygidial characters with those of *Platycalymene* and *Dionide*. While these single characters of *Rorringtonia* are comparable with those of the other three genera, the total aspect of the genus is rather different from any of them. Its overall sum of characters, including the forwardly tapering glabella, the preglabellar field, the opisthoparian suture and the number of thoracic segments (ten) are strongly suggestive of proetids, to which family it is here assigned. The type of glabella and the large number of pygidial pleural ribs and axial rings are comparable with the type species of *Pseudoproetus* Poulsen, *P. regalis* Poulsen, 1934, from the Llandovery of Greenland (see Moore 1959, p. O397, figs. 302, 2a–c) and the only major differences between *Rorringtonia* and *Pseudoproetus* are the more vaulted exoskeleton, the occipital ring of constant width (exsag.) and with lobes, and the path of the posterior branches of the facial sutures.

These differences are small, and there is a strong argument for considering the two genera con-generic, but until better preserved, uncrushed material of *Rorringtonia* is forthcoming, they are retained as separate genera.

Analocaspis (type species *A. ursina* Owens, 1970, p. 327, figs. 8A–H) from the Ordovician Chasmops Series of the Oslo district, Norway, bears some resemblance to *Rorringtonia*, particularly the incised glabellar furrows, position of the eye and the path of the posterior branch of the facial suture. It differs, however, in the structure of the preglabellar area and in the far smaller number of pygidial axial rings and pleural ribs. *Analocaspis* may be related to *Rorringtonia* and *Pseudoproetus*, and all three are difficult to relate to other Proetidae. Until more is known about them, all these genera are considered of uncertain subfamily.

Whittard's (1966, p. 292) diagnosis of *Rorringtonia* has been emended to incorporate characters exhibited by additional species assigned to the genus here.

1. **Rorringtonia flabelliformis** Whittard, 1966   Pl. 15, figs. 7, 8

1966   *Rorringtonia flabelliforme* Whittard, p. 292, pl. 50, figs. 8, 9.
1970   *Rorringtonia flabelliforme* Whittard; Owens, p. 328.

Type specimens. Holotype, GSM 102447, Pl. 15, fig. 8; external mould, figured Whittard 1966, pl. 50, fig. 8; from Rorrington Beds (Ordovician, Caradoc Series) 70 yd at 340° from bridge over Grey Grass Dingle, Rorrington, Shropshire (SJ 2989 0067). Paratype, GSM 102448, Pl. 15, fig. 7; partially exfoliated cranidium, figured Whittard 1966, pl. 50, fig. 9; from Rorrington Beds, tributary to Lower Wood Brook, 300 yd E of Desert, near Rorrington, Shropshire (SJ 3082 0166). These are the only known specimens of the species.

Diagnosis. Preglabellar field approximately $\frac{1}{2}$ sagittal length of glabella; anterior branches of facial sutures weakly divergent; eye ridge present: palpebral lobe about $\frac{1}{4}$ sagittal length of glabella; posterior portion of fixed cheek large, over $\frac{1}{2}$ basal glabellar width.

Description. See Whittard 1966, p. 292.

Measurements.

Cranidia	A	A_1	A_2+A_3	A_4	K	$\delta-\delta$	
GSM 102447 (E)	4·2	2·6	1·1	0·5	2·7	—	HOLOTYPE
GSM 102448 (E/I)	4·8	2·9	1·5	0·4	3·3	4·5	PARATYPE

Pygidium	Z	Y	W	X	
GSM 102447 (E)	(2·8)	(2·4)	(4·2)	(0·9)	HOLOTYPE

2. **Rorringtonia vetula** (Reed, 1935)   Pl. 15, figs. 9, 10

1935   *Proetus vetulus* Reed, p. 41, pl. 2, fig. 16.

Holotype. By monotypy; BM In36959, Pl. 15, figs. 9a, b; internal mould of cranidium, with counterpart external mould, figured Reed 1935, pl. 2, fig. 16; from Balclatchie Group (Caradoc Series), Balclatchie, near Girvan, Ayrshire (NX 256 968).

Material, horizon and locality. Two imperfect internal moulds of cranidia, BM In37491, BM In37009, both from the type locality.

Diagnosis. Preglabellar field about $\frac{1}{3}$ sagittal length of glabella; anterior branches of facial sutures almost parallel; no eye ridge; palpebral lobe $\frac{2}{5}$ sagittal length of glabella; posterior portion of fixed cheek small, $\frac{1}{5}$ basal glabellar width.

Description. Cranidium weakly vaulted, with palpebral width and sagittal length almost equal. Glabella somewhat broader than long, weakly convex in lateral and longitudinal profiles, tapering gently forwards with a well rounded frontal lobe, and defined by narrow, distinct conjoined axial and preglabellar furrows. Three pairs of incised lateral glabellar furrows. 1p: abaxial end about $\frac{2}{5}$ of glabella length from posterior; from this point 1p deepens gradually abaxially, directed weakly backwards until about $\frac{2}{3}$ of way towards sagittal line where it turns strongly backwards, shallowing rapidly and not reaching occipital furrow. There is a very short anterior branch, joining posterior branch where it turns sharply backwards. 2p: slightly posterior to γ, a little shorter than abaxial part of 1p, and much shallower, directed backwards at about same angle. 3p: a short distance

6

behind anterolateral corner of glabella, directed backwards at about same angle as 2p and slightly shorter and shallower.

Occipital furrow arched forwards sagittally, where it is narrow, widening a little abaxially, before narrowing again at extreme abaxial ends. Occipital ring about $\frac{1}{4}$ sagittal length of glabella, and transversely about the same width. Widest sagittally, narrowing rapidly laterally. Small median tubercle present, but no lateral lobes.

Preglabellar field about $\frac{1}{3}$ sagittal length of glabella and is convex in profile, rather steeply declined to shallow anterior border furrow which defines narrow, convex anterior border. Anterior branches of facial sutures almost parallel. γ a short distance out from axial furrow. No eye ridge. Palpebral lobe rather narrow and subcrescentic, about $\frac{2}{5}$ sagittal length of glabella. Posterior branches define small triangulate posterior portion of fixed cheek, ϵ and ξ a single angle. Sculpture apparently smooth.

Measurements.

Cranidia	A	A$_1$	A$_2$+A$_3$	A$_4$	K	δ–δ	
BM In36959 (I)	2·3	1·4	0·6	0·3	1·7	2·2	HOLOTYPE
BM In37491 (I)	—	(1·5)	0·4	—	1·5	—	
BM In37009 (I)	—	(1·4)	0·6	—	(1·3)	—	

Remarks. Comparison of the holotype of *Proetus vetulus* with that of *Rorringtonia flabelliformis* shows such similarity between the two species that *vetulus* should be transferred to *Rorringtonia*. *R. vetula* is known only from cranidia, and these differ from that of *R. flabelliformis* in the shorter preglabellar field, larger palpebral lobe, lack of eye ridges and in the smaller posterior portion of the fixed cheek.

3. **Rorringtonia sp. 1**   Not figured

Material, horizon and locality. BM It3022, internal mould of cranidium with counterpart external mould from black shales (Upper Llandeilo Series, low in *N. gracilis* Zone), Gwern-yfed-fâch, $\frac{1}{2}$ mile SE of Builth Road station, Radnorshire (SO 030 526).

Remarks. Dr C. P. Hughes (personal communication, 1970) has drawn my attention to a single specimen of *Rorringtonia* from the Upper Llandeilo of the Builth area, and has kindly supplied me with photographs. The cranidium is poorly preserved, but closely resembles *R. flabelliformis*, differing in the more elongated glabella, unforked 1p furrows and in (apparently) lacking the eye ridges. Until more material is available, it is not possible to decide whether this specimen really is specifically distinct from *R. flabelliformis*, and it is here referred to *Rorringtonia* sp. 1. Dr Hughes intends to describe and figure the specimen in a forthcoming part of the *Ordovician Trilobite faunas of the Builth-Llandrindod inlier, Central Wales.* [*Bull. Br. Mus. nat. Hist.* (Geol.)].

Indeterminate PROETIDAE

A small number of fragmentary or poorly preserved specimens, which are not readily referable to any of the species described in this monograph, are recorded here for completeness.

Proetid 1   Pl. 15, fig. 11
1950   Proetid cephalon; Whittington, p. 33.

Material, horizon and locality. SM A38941, a poorly preserved internal mould of a cephalon from Slade Beds (Ashgill Series), Upper Slade, Haverfordwest, Pembrokeshire.

Description. Cephalon with broad, flattened border; elongated, coniform glabella; very short (sag.) preglabellar field; long crescentic eye; strongly divergent anterior branches of facial sutures; broad-based genal spine; occipital ring apparently without lobes.

Remarks. The characters exhibited by this specimen are difficult to equate with those of any established proetid genus. The glabellar shape and strongly divergent anterior branches of the facial sutures are reminiscent of *Astroproetus reedi* (see Pl. 10, figs. 18–21), but the very short (sag.) preglabellar field, the wide cephalic border and the apparent lack of lateral occipital lobes are not typical of *Astroproetus*; the first and last of these characters are found in some *Proetus* species, e.g. *P. berwynensis* (Pl. 1, fig. 1). No other proetids have been recorded from the Slade Beds.

Proetid 2   Pl. 15, fig. 12

Material, horizon and locality. OUM C14616, free cheek from Pentamerus Beds (Llandovery Series), 325 yd ESE of Merrishaw, near Harley, Shropshire (SJ 5852 0065) [ZCM loc. 54].

Remarks. This incomplete free cheek has a pitted sculpture of a type not seen in any other Llandovery proetids, but comparable with such species as *Proetus* (*Lacunoporaspis*) *confossus* and *P.*(*L.*) *obconicus* from the Wenlock and Ludlow respectively (see Pl. 4, figs. 4, 8; Pl. 5, figs. 1a–c).

Proetid 3   Pl. 15, figs. 13–15

Material, horizons and localities. OUM C14141, OUM C14153, free cheeks from 700 yd SW of church at Hughley, Shropshire (SO 5605 9747) [ZCM loc. 56]; OUM C13173, pygidium from 400 yd ENE of Wall-under-Heywood, Shropshire (SO 5120 9276) [ZCM loc. 58]. All from Llandovery, Telychian Stage.

Remarks. These specimens appear to belong to one species; both the pygidium (Pl. 15, fig. 13) and one of the free cheeks (Pl. 15, fig. 15a) show a broad, weakly convex doublure. One free cheek (Pl. 15, figs. 15, 15a) shows a distinct eye socle, like that of some *Decoroproetus* species, but the type of pleural ribs on the pygidium do not appear to be *Decoroproetus*-like, although the preservation is poor.

SPECIES INCORRECTLY ASSIGNED TO PROETID GENERA.

Examination of the type material of the three following species has shown them not to be proetids:

Proetus brachypygus Marr & Nicholson, 1888, p. 725, pl. 16, figs. 18, 19; from Upper Skelgill Beds (Llandovery, *Ampyx aloniensis* Bed), Lake Windermere district, Westmorland. Assigned here to *Otarion*.

Cyphoproetus newlandensis Begg, 1950, p. 287, pl. 14, fig. 7; from Saugh Hill Sandstones (Llandovery), Newlands, near Girvan, Ayrshire. Assigned here to *Otarion*.

Proetus (*Astroproetus*) *whittardi* Begg, 1939, p. 376, pl. 6, fig. 3; from Upper Drummuck Group, Starfish Bed no. 2, (Ashgill, Rawtheyan Stage), Lady Burn, near Girvan. Assigned here to *Panarchaeogonus*.

PHYLOGENY OF ORDOVICIAN AND SILURIAN PROETIDAE

INTRODUCTION.

The following remarks are based mainly on information which is derived from Proetidae from Britain (described herein) and from Scandinavia (Owens 1970, 1973), which are taken to provide a representative sample of the earlier members of the family. Information available from elsewhere is incorporated, but as many of the published illustrations and descriptions are old and inadequate, and much of the material has not been examined at first hand, major conclusions have not been drawn from those sources. The available evidence does, however, appear to fit into the picture derived from the British and Scandinavian material.

ORIGINS AND EARLY HISTORY OF THE PROETIDAE.

As regards the origin of the Proetidae, many later Cambrian and early Ordovician ptychopariid trilobites have a generalized 'proetid' morphology—e.g. *Elrathia* Resser, 1937, *Weeksina* Resser, 1935 and *Hystricurus* Raymond, 1913 (see Moore 1959, p. O241, fig. 179, 1; p. O281, fig. 207 and p. O277, figs. 204, 4a–c respectively), especially in the forwardly tapering glabella, opisthoparian suture and well developed preglabellar field. Whittington (1966, p. 709) has already suggested a relationship between certain hystricurines and Proetacea such as *Otarion*, *Dimeropyge* and *Phaseolops*, and genera classified with the former (see Moore 1959, p. O278) do seem very plausible candidates for ancestors of the latter. For example, distinct similarities can be seen between *Otarion* and *Psalikilus* and between *Dimeropyge* and *Psalikilopsis*. *Hystricurus* itself is of particular interest as it has a small, triangular, rostral plate whose connective sutures converge backwards, just like the majority of proetids. *Hystricurus* is broadly similar to *Decoroproetus*, although its granular surface sculpture contrasts with the striated one typical of the latter.

Proetid trilobites do not occur in any abundance until the Llandeilo, in which the earliest species of the tropidocoryphine *Decoroproetus* are found. Dean (1966, p. 338, pl. 15, figs. 1, 7) described and figured a cranidium from the Arenig (*extensus* Zone) of the Montagne Noire, S France, which he supposed to be a proetid. The overall morphology (forward-tapering glabella, preglabellar field, opisthoparian suture) is certainly proetid-like, although the lack of the occipital furrow is atypical of proetids. The affinities of this specimen are difficult to assess without more material. The earliest definite proetid known to me comes from the Tourmakeady Limestone (Arenig, *hirundo* Zone) of County Mayo, western Ireland, and is represented by a few cranidia collected recently by Dr. R. A. Fortey, who has kindly allowed me to examine them. These cranidia are *Decoroproetus*-like in general morphology, but have a finely granular sculpture. In many ways, they are almost intermediate between *Hystricurus* and *Decoroproetus*, and thus add weight to the possible existence of such an evolutionary lineage.

Phaselops sepositus (Whittington, 1963, p. 37, pl. 4, figs. 11–13; pl. 5, figs. 1–3) from beds equivalent to the Whiterock Stage (correlated with the Llanvirn) of western Newfoundland is the only other definite pre-Llandeilo proetid known to me. It is a tropidocoryphine, and appears to be a small, specialised offshoot from the main evolutionary line.

The early Ordovician is clearly the critical period for early proetid evolution, although unfortunately the record to date is scant. It is now possible, however, to draw a far more detailed picture of the post-Llanvirn history of the Proetidae than has hitherto been possible, and each subfamily is examined in turn below.

PHYLOGENIES OF SUBFAMILIES.

Tropidocoryphinae. The ancestry of this subfamily possibly lies in *Hystricurus*, and *Decoroproetus* or *Decoroproetus*-like genera are considered to form its main root-stock. The earliest species definitely referable to *Decoroproetus* occur in the Llandeilo. In the Caradoc and Ashgill the genus becomes widespread, represented by many species in Britain (described herein) and Scandinavia (Owens 1970, 1973), although it appears to be absent from Bohemia and the Mediterranean region at this time. *Decoroproetus* persists into the Silurian and Lower Devonian, but present evidence suggests that it became far more restricted in its distribution after the Ordovician. It is absent, for example, from the carbonates of Wenlock age well developed in the Welsh Borderland and Gotland, and appears to be confined to dark, fine-grained limestones and shales occurring in Britain in the Long Mountain and the Sedbergh area, and in Scania, Sweden; in the Holy Cross Mountains, Poland, and in Bohemia. In the Lower Devonian it has been recorded in Bohemia (Přibyl 1949), Morocco (Alberti 1969) and in Nevada, U.S.A. (Haas 1969).

Throughout its long history, *Decoroproetus* changes little in its basic morphology, and it is difficult to pick out any general phylogenetic trends. Several different tropidocoryphine lines can be traced back to *Decoroproetus*, and appear to have arisen from it by iterative evolution.

Stenoblepharum Owens, 1973, appears sometime during the Caradoc, and its earliest known species, *S. strasburgense* (Cooper 1953, p. 19, pl. 1, figs. 15–19) occurs in the Edinburg Formation, Virginia, U.S.A. The genus may have arisen from *Decoroproetus* by reduction in number and modification of the pygidial pleural ribs, by shortening of the preglabellar field and general narrowing and increase of vaulting of the exoskeleton. All known species of *Stenoblepharum* are found in carbonate deposits, and are particularly characteristic of late Ordovician (e.g. Boda and Kildare) reef limestones, in which they commonly occur in great abundance.

In the late Ashgill, the short-lived *Paraproetus* appears, and has so far only been recorded from beds of this age in certain parts of the British Isles and in the Holy Cross Mountains, Poland. *Paraproetus* is characterized by reduced eyes and by pygidial pleural ribs which are almost intermediate in their structure between those of *Decoroproetus* and those of the Cornuproetinae, and it may thus be associated with the ancestry of the latter.

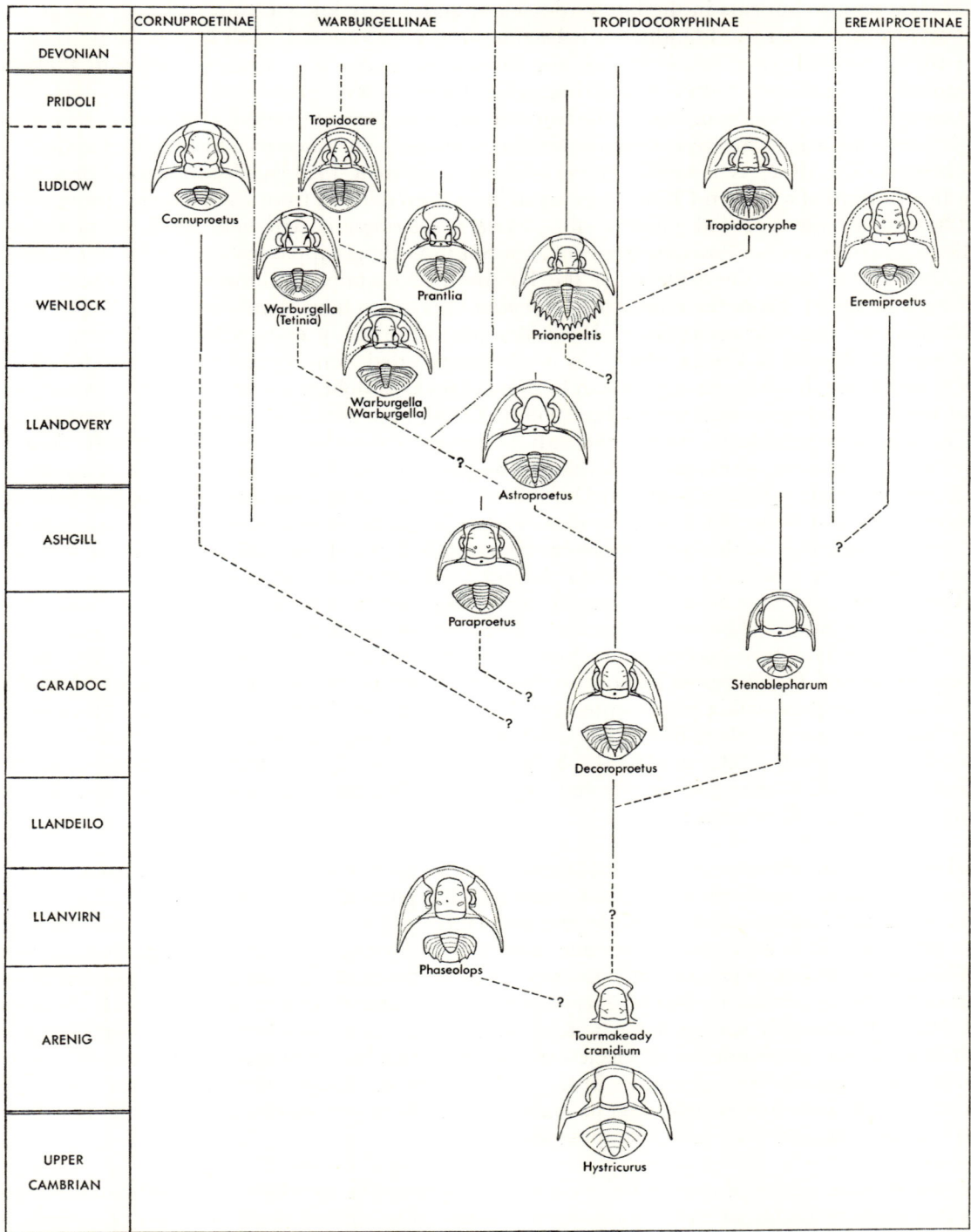

TEXT-FIG. 11. Phylogeny of earlier genera of the Tropidocoryphinae, Warburgellinae, Cornuproetinae and Eremiproetinae. The cornuproetines *Lepidoproetus* and *Lodenicia* are omitted, as is the probable tropidocoryphine *Lati-*
proetus. The two former genera probably have their roots in *Cornuproetus*, in the Wenlock or Ludlow, whilst the
latter was probably derived from *Decoroproetus* in Silurian times. Solid lines represent known ranges of genera,
broken ones their possible derivation.

Astroproetus also appears in the late Ordovician, and is only known from rocks of this age from the Girvan district. In the Llandovery it becomes more widespread, with species being recorded from the Girvan district; South Wales; the Oslo district, Norway, and the Siberian Platform. *Astroproetus* has not been found in post-Llandovery strata. While the cephalon is similar to that of *Decoroproetus*, the pygidium has a rib-structure more or less intermediate between *Decoroproetus* and *Warburgella*, and *Astroproetus* seems to be associated with the ancestry of the former genus.

Latiproetus Lu, 1962, (type species *Proetus latilimbatus* Grabau, 1924 [see Lu 1962, pl. 1, figs. 3–6]), from the Silurian of southern China, appears to have been derived from *Decoroproetus* through some modification of the pygidial pleural ribs. '*Proetus*' *bowningensis* Mitchell, 1887, from the late Ludlow—early Devonian Bowning Series of New South Wales, is apparently referable to *Latiproetus*, as is '*Proetus*' sp. from the Silurian of the Klamath Mountains, California (see Churkin 1961, p. 173, pl. 35, fig. 9). From these records, *Latiproetus* appears to have flourished in the Pacific region in Silurian and early Devonian times. It is unknown in Europe or eastern North America.

Prionopeltis Hawle & Corda, 1847, has not been recorded from the Silurian of the British Isles or of Scandinavia, but is quite common in Ludlovian rocks in Bohemia and in the Harz Mountains, Germany. It is characterized (see Moore 1959, p. O396, figs. 301, 2) by weakly developed 1p lobes and by a spinose pygidium, and is the only tropidocoryphine with the latter feature. Pygidial spines are also present, presumably developed by parallel evolution, in *Phaetonellus*, a member of the Eremiproetinae which is found in Devonian strata of the Hercynian facies. It has been suggested elsewhere (Owens, 1973, p. 146) that *Prionopeltis* might be related to *Decoroproetus* through a late Ordovician species such as *D. campanulatus* Owens (1973, p. 141, figs. 5D-K), which has a pygidial pleural rib-structure similar to species of *Prionopeltis* but lacks the marginal spines.

There appears to have been a late tropidocoryphine evolutionary burst during later Silurian and early Devonian times, when such genera as *Tropidocoryphe* and *Astycoryphe* made their appearance. Many of these late tropidocoryphines have a tropidium, or a series of tropidia. The earliest representatives of *Tropidocoryphe* I know are from the Liteň Formation (late Wenlock) of Svatý Jan pod Skalou, Czechoslovakia, represented by three pygidia in the British Museum (Natural History) [all under one number, BM 42384], belonging to a new, undescribed species. '*Proetus*' *rattei* Etheridge & Mitchell (1892, p. 316, pl. 25, figs. 2a, 2c, 2 [*non* 2b, =*Scharyia*?]), from the Ludlow of New South Wales, may also be referred to *Tropidocoryphe*. Hitherto, *Tropidocoryphe* has not been recorded from pre-Devonian rocks.

Warburgellinae. Members of this subfamily, which appears to have originated from an *Astroproetus*-like ancestor, are typified by flat-topped pygidial pleural rib-structure, deep 1p furrows and by a rostral plate with backwardly diverging connective sutures, the latter a condition not found in other proetids. There is evidence of parallel evolution occurring between warburgellines and other subfamilies, for *Warburgella* is superficially similar to the proetine *Cyphoproetus*, while the cephalon of *Tropidocare* is very similar to that of *Tropidocorpyhe*. Like later tropidocoryphines, both *Tropidocare* and many *Warburgella* species have a tropidium. The third member of the Warburgellinae, *Prantlia*, lacks the tropidium in all species, but whether this lack is primary or secondary is uncertain. All three genera become extinct in the early Devonian. *Koneprusites* Přibyl, 1964, has a pygidial pleural rib-structure which suggests that it might be a late warburgelline, although Alberti (1969, p. 246) placed it in the Cornuproetinae.

Cornuproetinae. The type of pygidial pleural ribbing found in members of this subfamily suggests that its origins probably lie in the Tropidocoryphinae. The earliest described Cornuproetinae are of Wenlock age, but certain specimens from the late Ordovician Porkuni Stage of Estonia in the collections of the Naturhistoriska Riksmuseet, Stockholm are, in my opinion, cornuproetines— witness the structure of their pygidial pleural ribs which is imbricate and the overall morphology of the cephalon. Hence the subfamily probably has its origins in the later Ordovician, possibly to be traced to a *Decoroproetus* species such as *D. furubergensis* (Owens 1970, p. 312, figs. 5G, J, K;

figs. 6G, H, J), which has a pygidial pleural rib-structure similar to that of certain cornuproetines (see Owens *op. cit.*, p. 315).

Cornuproetinae are rare in the Silurian of Britain, and are unknown in Scandinavia, but are common in the later Silurian of Bohemia. The subfamily reaches its maximum diversification in the Devonian, and has recently been discussed in detail by Alberti (1969, p. 116), who has also proposed a phylogeny for the various subgenera of *Cornuproetus*. Although not discussing post-Silurian Cornuproetinae in detail here, it is of interest to note that the type species of *Cornuproetus*, *C. cornutus* (Goldfuss, 1843) has a preannulus (see R. & E. Richter 1956, pl. 1, fig. 4; pl. 2, fig. 9), although at least the Silurian species currently assigned to this genus have not. It is thus possible that the preannulus was developed independently in some later species of *Cornuproetus*, although it may reflect some relationship with the Proetinae.

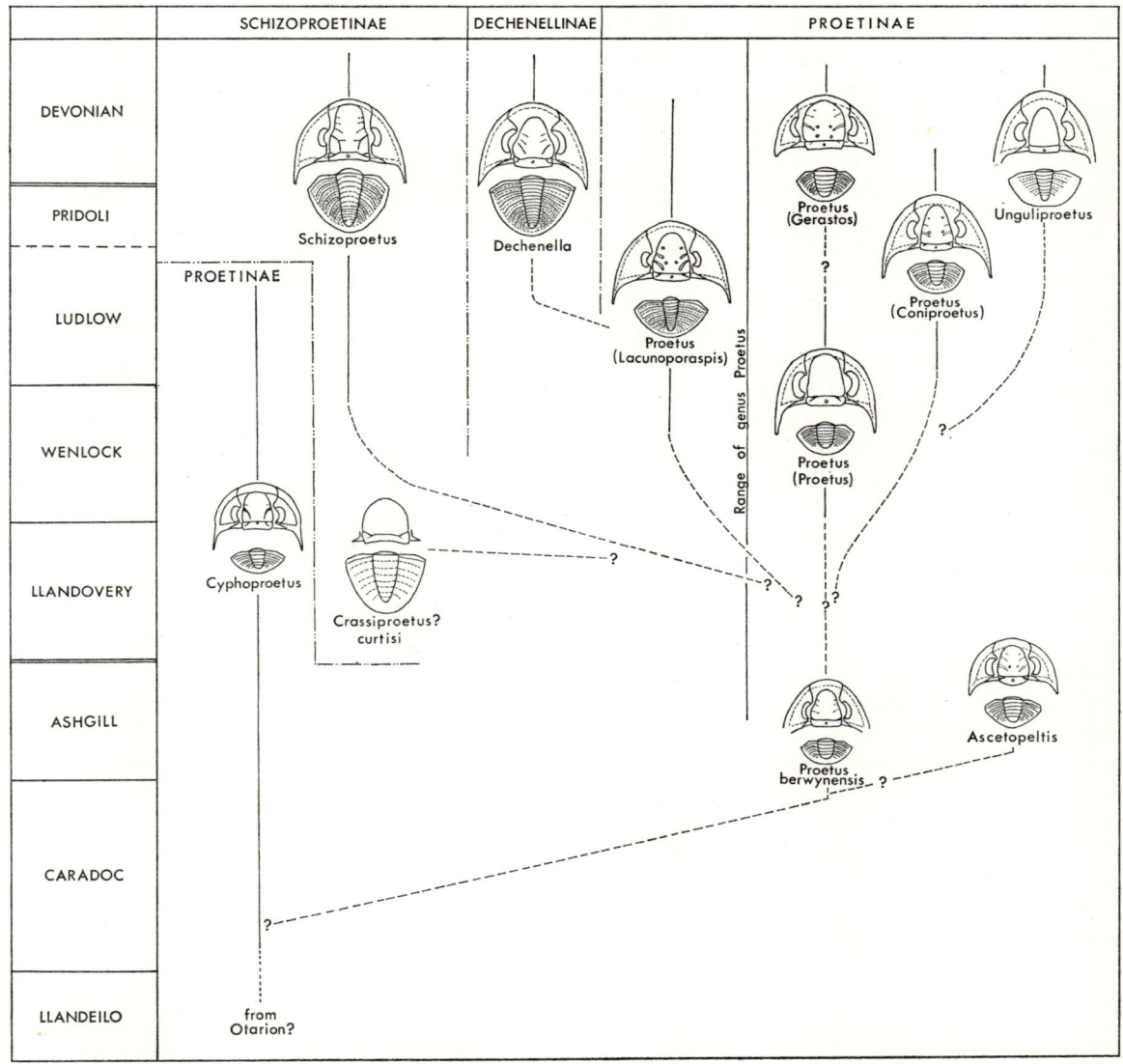

TEXT-FIG. 12. Phylogeny of earlier genera of the Proetinae, Dechenellinae and Schizoproetinae. Solid lines represent known ranges of genera, broken ones their possible derivation.

Eremiproetinae. As currently conceived, this subfamily comprises two genera, *Eremiproetus* and *Phaetonellus.* The earliest record of the former (and of the subfamily) is in the late Ordovician (Owens 1973, p. 167), where it is represented by one species, *E. agellus* Owens, 1973. From the Silurian only one species of *Eremiproetus, E. senex* Alberti, 1967, has been described. The genus reaches its greatest abundance and diversification in the Devonian, from which many species have been described (see Alberti 1969). The pygidial pleural rib-structure of *Eremiproetus* is imbricate, and similar to that of *Decoroproetus,* while the cephalon of *E. agellus* is similar to those of *Stenoblepharum* species. *Eremiproetus* may, therefore, have its origins in *Decoroproetus* or *Stenoblepharum.*

Proetinae. Cyphoproetus apparently forms the ancestral stock of this subfamily. It is found in small numbers in the mid- and late Ordovician and throughout the Silurian, towards the end of which it became extinct. The principal evolutionary trend in *Cyphoproetus* is the development of the sagittally widened anterior border and the incurving of the lateral margin of the cephalon at the base of the genal spine. These features characterize mid-Silurian species such as *C. depressus* and *C. strabismus.* The surface sculpture, where observed, is striated in earlier species (e.g. *C. facetus*), but granular in later ones (e.g. *C. depressus*).

The genus *Proetus* might easily have developed from *Cyphoproetus* simply by suppression of the deep 1p furrows—other exoskeletal characters of the two genera are on the whole very similar. The earliest *Proetus* species are found in the Ashgill, and on these the preannulus is well developed, e.g. in *P. berwynensis* (Pl. 1, fig. 1). The problems of differentiating different subgenera are outlined under *Proetus* (p. 8).

Lacunoporaspis, here considered to be a subgenus of *Proetus* on phylogenetic and morphological grounds, first appears in the Wenlock and persists to the Middle Devonian. It is almost certainly connected with the ancestry of the Dechenellinae, and such species as *L. norrisi* (Ormiston, 1971, p. 31, pl. 3, figs. 1–7, 9–15) are alomst intermediate in their morphology between *Lacunoporaspis* and *Dechenella.* Ormiston (1967, p. 70) has recently suggested ancestry of the Dechenellinae in *Proetus (Longiproetus),* but *Lacunoporaspis* would appear to be a much more probable ancestor.

Schizoproetus has been classified with the Dechenellinae (e.g. by Richter, Richter & Struve *in* Moore 1959, p. O389), but Yolkin (1968, p. 43) has recently proposed the subfamily Schizoproetinae to accommodate it. This subfamily is under study by the author who proposes to slightly modify Yolkin's concept of it. The earliest *Schizoproetus* species are found in the Wenlock, and their ancestry, to judge from pygidial and cranidial characters, is probably associated with *Proetus (Lacunoporaspis).*

Proetidae incertae sedis. There are several genera in the Ordovician and Silurian which, although readily assigned to the Proetidae as here conceived, are difficult to associate with any of the established subfamilies discussed above.

One group of these genera includes *Rorringtonia, Pseudoproetus* and *Analocaspis,* and these three might form a distinct taxonomic unit, although there is not enough information available at this stage to propose a new subfamily for them. Their origins possibly lie in the Tremadoc genus *Protarchaegonus* (Sdzuy, 1955, p. 40, pl. 7, figs. 2–7), and they are difficult to relate to either the Tropidocoryphinae or the Proetinae.

Xenocybe Owens (1973, p. 173, figs. 14D, F–J) is difficult to relate to any other genus, especially as the pygidium is unknown.

The genera *Scharyia* Přibyl, 1946, *Panarchaeogonus* Öpik, 1937, and *Isbergia* Warburg, 1925, which have been included in the Proetidae in the past, are more closely related to the otarionids.

CONCLUSIONS.

Two major groups may be recognized in the subfamilies considered above. The Warburgellinae, Eremiproetinae and probably the Cornuproetinae are likely to have their origins in the Tropidocoryphinae. The history of the last-named extends back with certainty to the Llanvirn, with *Phaseolops* as the earliest representative, and possibly to the Arenig Tourmakeady cranidia

(Text-fig. 11). The earliest proetine, *Cyphoproetus*, occurring in the Caradoc, bears no close resemblance to any tropidocoryphine (the only contemporaneous proetid subfamily known), nor does it particularly resemble any of the Proetidae *incertae sedis*. The otarionid-like appearance of *Cyphoproetus* is possibly indicative of a closer association with *Otarion*. If this is so, then the Proetinae and their descendants are likely to have an origin quite separate from the Tropidocorphinae and their descendants. Thus the Proetidae, as currently conceived (e.g. Moore 1959, pp. O382–98), comprises two separate phylogenetic stocks. If this could be demonstrated with more certainty in the future there would be a good case for considering two independent families to be represented, although such a division at this stage would be premature.

DISTRIBUTION OF PROETIDAE IN THE BRITISH ISLES

ORDOVICIAN.

Llandeilo Series. The subspecies *Decoroproetus fearnsidesi pristinus* appears to range throughout the Llandeilo Series, and has been found in the Berwyn Hills and in the Llandeilo region in small numbers, but has not been recorded from the Builth Inlier. From the latter a single cranidium of *Rorringtonia* sp. has been recorded.

Caradoc Series. There is differentiation of the proetid faunas, like other faunas, between the Girvan and the Anglo-Welsh areas.

In the Anglo-Welsh area, *Decoroproetus fearnsidesi fearnsidesi* is the characteristic proetid of the Costonian and Harnagian Stages in Shropshire, and in 1971 was found by Dr. R. Addison in uppermost Llandeilo or basal Caradoc deposits in the Narberth area, Pembrokeshire. It has not been found in strata in North Wales correlated with the Costonian and Harnagian. *Rorringtonia flabelliformis* occurs rarely in basal Caradoc strata in the Shelve Inlier.

In the Soudleyan, *D. fearnsidesi* is replaced by a similar species of *Decoroproetus*, *D. calvus*. This species has been widely recorded from strata referred to the Soudleyan and Longvillian Stages in South Shropshire, the Shelve Inlier, the Berwyn Hills, the Bala district, the Dolwyddelan district and the Cross Fell Inlier. It also possibly occurs in later Caradoc strata, but the record is uncertain.

Two proetid species, *Decoroproetus jamesoni* and *Rorringtonia vetula*, are found in the Caradoc Balclatchie Group of the Girvan district. *D. jamesoni?* is recorded from the Upper Stinchar Limestone and from the *confinis* Flags (see p. 47). The Caradoc Craighead (Kiln) Mudstones have yielded the earliest known species of *Cyphoproetus*, *C. facetus*. The genus has not been recorded from the Ordovician of the Anglo-Welsh area.

Ashgill Series. In the Anglo-Welsh area a succession of proetid species is found in Ashgill strata. *Decoroproetus piriceps* is characteristic of the Cautleyan Stage, and possibly also occurs in the higher Pusgillian. It has been recorded from the Lake District, the Cautley and the Llandeilo districts. In the latter it is found in the Birdshill Limestone, probably of Pusgillian age, and also in the Crûg Limestone, whose age is somewhat conjectural, and may even be Marshbrookian (see p. 48). In Pembrokeshire, in various strata considered to be Cautleyan the earliest known form of *Proetus* (s.l.), *P.* cf. *berwynensis*, occurs. *P. berwynensis* itself is from beds of mid-Ashgill age from Cynwyd, N Wales.

In the Rawtheyan Stage, *D. piriceps* is replaced by *D. papyraceus*. This species has been recorded from the Craven inliers, NW Yorkshire, and probably also occurs at Cautley. In the topmost Rawtheyan of the Lleyn Peninsula and of the Coniston area, Lancashire, *Paraproetus girvanensis* is found in small numbers. This species is very common in the late Rawtheyan Upper Drummuck Group of the Girvan area where it is accompanied by a diverse proetid fauna including *Astroproetus reedi*, *Decoroproetus asellus* and *Cyphoproetus rotundatus*. None of these other species has been found in the Anglo-Welsh area.

The Keisley Limestone reef and the possibly equivalent limestone of the Horton-in-Ribblesdale neptunean dyke have yielded a few specimens referred to here as *Decoroproetus* cf. *subornatus*. *D. subornatus* occurs in the Whitehead Formation of Quebec. The proetid fauna of the Kildare Limestone of the Chair of Kildare, eastern Ireland, like that of the Ashgill Boda Limestone of Dalarne, Sweden, is dominated by species of *Stenoblepharum* Owens, 1973, a genus which has not yet been reported from the Keisley Limestone. Dr. W. T. Dean proposes to describe the Kildare proetids in a forthcoming part of his monograph of the trilobites of the Chair of Kildare Limestone. The proetids from the Boda Limestone have been described by Owens (1973).

In beds belonging to the highest part of the Hirnantian Stage or to the basal Llandovery (see p. 51, under *D.* cf. *evexus*), a species referred to here as *Decoroproetus* cf. *evexus* occurs, which might be conspecific with *D. evexus*, which occurs in beds high in the Harju Series (=Ashgill), equated with the high Rawtheyan and Hirnantian Stages (see Owens 1973, p. 146).

SILURIAN.

Llandovery Series. As in the Ordovician, the proetid faunas from Girvan and from the Anglo-Welsh area are distinct, with the former showing closest affinities with Norwegian faunas (as yet undescribed).

The diversity of proetids found in the Llandovery of the Anglo-Welsh area is much greater than in the Ordovician, and species of *Proetus* (s.l.), *Cyphoproetus*, *Crassiproetus*?, *Decoroproetus*, *Astroproetus* and *Warburgella* occur, mostly in beds belonging to the Fronian and Telychian Stages. Of these, *Proetus* (s.l.) *latifrons* and *Cyphoproetus binodosus* are also found in the Wenlock.

In contrast to those of the Anglo-Welsh area, the faunas of the Girvan district include only species of *Cyphoproetus* and *Astroproetus*. The one species of the former, *C. externus*, occurs in beds correlated with the Rhuddanian and Idwian Stages, while three species of the latter, *A. scoticus*, *A. interjectus* and *A. pseudolatifrons* are found in the Rhuddanian, Idwian and Fronian Stages respectively.

Wenlock Series. In the basal Wenlock of the Anglo-Welsh area, *C. binodosus* survives from the Llandovery. In the early Wenlock carbonate developments of the Woolhope and Dolyhir Limestones, *Warburgella scutterdinensis* and *Cornuproetus peraticus* occur respectively. There are many specimens from old museum collections simply labelled 'Wenlock Shale', and among these are the species *Proetus* (*Proetus*) *concinnus*, *P.* (s.l.) *falcatus* and *Cyphoproetus depressus*. Accurately localized specimens of *P.* (*P.*) *concinnus* come from the Wenlock Limestone or high Wenlock Shale, while *C. depressus* is recorded from late Wenlock siltstones of the Penylan Inlier, Cardiff. In the Wenlock Limestone of Wenlock Edge, only *P.* (*P.*) *concinnus* and rare *Warburgella stokesii* occur, but the approximately coeval Nodular Beds at Dudley have yielded a much richer fauna, and besides the two latter species, *Proetus* (*Lacunoporaspis*) *confossus* and *Schizoproetus* aff. *delicatus* are recorded; *P.* (*L.*) *confossus* is dominant. The remainder of the trilobite fauna in the Nodular Beds at Dudley is far more diverse than at Wenlock Edge.

In the Long Mountain and in the Cautley area, late Wenlock and early Ludlow shales and dark, fine-grained limestones contain *Decoroproetus scrobiculatus* (in both areas) and *Cyphoproetus strabismus* (Long Mountain only). These species are not found in the typical carbonate developments of late Wenlock age, and the same, or closely related species, occur in similar, approximately contemporaneous, beds in Scania, southern Poland and Bohemia.

Ludlow Series. Proetus (*Proetus*) *concinnus* is found in the Lower Elton Beds, but not in later strata. The remainder of the British Ludlow contains only three proetid species: *Proetus astringens* occurs in small numbers in the Upper Elton and Lower Bringewood Beds in the Welsh Borderland; *Proetus* (*Lacunoporaspis*) *obconicus* occurs in the Upper Bringewood and Leintwardine Beds, reaching its acme at about the boundary of the Lower and Upper Leintwardine Beds, and has been found in the Welsh Borderland and in the Kendal area. *Warburgella* (*Tetinia*) *ludlowensis* has a similar distribution to *P.* (*L.*) *obconicus*, but is found in small numbers only.

REFERENCES

ALBERTI, G. K. B. 1963. Zur Kenntnis rheinisch-herzynischer Mischfaunen (Trilobiten) im Unterdevon. *Mitt. Geol. Staat. Hamburg*, **32**, 148–59, pls. 15, 16.

—— 1964. Neue Trilobiten aus dem Marokkanischen und deutschen Unter- und Mitteldevon. *Senckenb. leth.*, **45**, 115–32, pls. 16, 17.

—— 1966. Über einige neue Trilobiten aus dem Silurium und Devon, besonders von Marokko. *Ibid.*, **47**, 111–21, pl. 14.

—— 1967. Neue obersilurische sowie unter- und mitteldevonische Trilobiten aus Marokko, Deutschland und einigen anderen europäischen Gebieten. 2. *Ibid.*, **48**, 481–509, pl. 1.

—— 1969. Trilobiten des jüngeren Siluriums sowie des Unter- und Mitteldevons. I. Mit Beiträgen zur Silur-Devon Stratigraphie einiger Gebiete Marokkos und Oberfrankens. *Abh. Senck. Nat. Gesell.*, **520**, 1–692, pls. 1–52.

—— 1970. Trilobiten des jüngeren Siluriums sowie des Unter- und Mitteldevons. II. *Ibid.*, **525**, 1–233, pls. 1–20.

ALTH, A. VON 1874. Über die paläozoischen Gebilde Podoliens und deren Versteinerungen. *Abh. K.-K. geol. Reichsanst*, **7**, 1–79, 5 pls.

ANGELIN, N. P. 1851. Palaeontologia Suecica. I: *Iconographia crustaceorum formationis transitionis.* Fasc. 1, 1–24, pls 1–24. Lund.

—— 1854. *Palaeontologia Scandinavica. I. Crustacea formationis transitionis.* Fasc. 2, I–IX, 21–92, pls. 25–41. Lund.

BALASHOVA, E. A. 1968. [*Trilobites of the Skala and Borshchev horizons in Podolia. In: Siluro-Devonian faunas of Podolia*], 95–124, 3 pls. Leningrad University [In Russian].

BANCROFT, B. B. 1949. Upper Ordovician trilobites of zonal value in south-east Shropshire. (Edited by A. LAMONT). *Proc. R. Soc.*, (*B*), **136**, 291–315, pls. 9–11.

BARRANDE, J. 1846. *Notice préliminaire sur le Système Silurien et les trilobites de Bohême.* vi+97 pp. Leipsic.

—— 1846a. *Nouveaux trilobites. Supplement á la notice préliminaire sur le Système Silurien et les trilobites de Bohême.* iv+40 pp. Prague.

—— 1852. *Système Silurien du centre de la Bohême. 1ère partie. Recherches paléontologiques*, vol. 1. *Crustacés, Trilobites.* xxx+935 pp. 51 pls. Prague & Paris.

—— 1872. *Système Silurien du centre de la Bohême. 1ère partie. Recherches paléontologiques.* Supplement au vol. 1. xxx+647 pp., 35 pls. Prague & Paris.

BASSETT, M. G. 1972. The articulate brachiopods from the Wenlock Series of the Welsh Borderland and South Wales. *Palaeontogr. Soc.* [*Monogr.*]: (2), 27–78, pls. 4–17.

BEGG, J. L. 1939. Some new species of Proetidae and Otarionidae from the Ashgillian of Girvan. *Geol. Mag.*, **76**, 372–82, pl. 6.

—— 1943. Hypostomes of some Girvan trilobites and their relationship to the cephala. *Ibid.*, **80**, 56–65, pl. 2.

—— 1947. Some new fossils from the Girvan district. *Trans. geol. Soc. Glasgow*, **21**, 29–47, pls. 2, 3.

—— 1950. New trilobites from Girvan. *Geol. Mag.*, **87**, 285–91, pl. 14.

—— 1951. Some new Girvan trilobites. *Trans. geol. Soc. Glasgow*, **21**, 362–70, pl. 1.

—— & REED, F. R. C. 1945. Two new trilobites from Girvan. *Ibid.*, **20**, 260–3, pl. 1.

BERGSTRÖM, S. 1964. Remarks on some Ordovician faunas from Wales. *Acta Univ. lund.* Sec. II. no. 3, 1–67.

BEYRICH, E. 1846. *Untersuchungen über Trilobiten.* Part 2. 37 pp., 4 pls. Berlin.

BURMEISTER, H. 1843. *Die Organisation der Trilobiten, aus ihren lebenden Verwandten entwickelt; nebst einer systematischen Übersicht aller zeither beschriebenen Arten.* 147 pp., 6 pls. Berlin.

—— 1846. *The organization of trilobites, deduced from their living affinities; with a systematic review of the species hitherto described.* (Edited from the German by BELL & FORBES). *Ray Soc. publs*, 136 pp., 6 pls. London.

CAMPBELL, K. S. W. 1967. Trilobites of the Henryhouse Formation (Silurian) in Oklahoma. *Bull. Okla. geol. Surv.*, **115**, 1–68, pls. 1–19.

CANTRILL, T. C., DIXON, E. E. L., THOMAS, H. H. & JONES, O. T. 1916. The geology of the South Wales Coalfield. Part XII. The country around Milford. *Mem. geol. Surv. U.K.*, vii+185 pp.

CAVET, P. & PILLET, J. 1958. Les trilobites des Calcaires à Polypiers Siliceux (Eifelian) du Synclinal de Villefranche-de-Confluent (Pyrénées-Orientales). *Bull. Soc. géol. Fr.*, (6), **8**, 21–37, pl. 3.

CHATTERTON, B. D. E. 1971. Taxonomy and ontogeny of Siluro-Devonian trilobites from near Yass, New South Wales. *Palaeontographica* (A), **137**, 1–108, pls. 1–24.

CHLUPÁČ, I. 1971. Some trilobites from the Silurian/Devonian boundary beds of Czechoslovakia. *Palaeontology*, **14**, 159–77, pls. 19–24.

CHURKIN, M. 1961. Silurian trilobites from the Klamath Mountains, California. *J. Paleont.*, **35**, 168–75, pls. 35–6.

COCKS, L. R. M., HOLLAND, C. H., RICKARDS, R. B. & STRACHAN, I. 1971. A correlation of Silurian rocks in the British Isles. *Jl geol. Soc.*, **127**, 103–36.

COOPER, B. N. 1953. Trilobites from the Lower Champlainian Formations of the Appalachian Valley. *Mem. geol. Soc. Am.*, no. 55, i–vi, 1–69, pls. 1–19.

COOPER, G. A. & KINDLE, C. H. 1936. New brachiopods and trilobites from the Upper Ordovician of Percé, Quebec. *J. Paleont.*, **10**, 348–72, pls. 51–53.

CURTIS, M. L. K. 1958. The Upper Llandovery trilobites of the Tortworth Inlier, Gloucestershire. *Palaeontology*, **1**, 139–46, pl. 29.

—— 1972. The Silurian rocks of the Tortworth Inlier, Gloucestershire. *Proc. Geol. Ass.* **83**, 1–35, pls. 1, 2.

DALMAN, J. W. 1827. Om palaederna, eller de så kallade trilobiterna. *K. svenska Vetensk-Akad. Handl.* (for 1826), 113–52, 226–94, pls. 1–6.

DEAN, W. T. 1958. The faunal succession in the Caradoc Series of South Shropshire. *Bull. Br. Mus. nat. Hist.* (Geol.), **3**, (6), 193–231, pls. 24–26.

—— 1959. The stratigraphy of the Caradoc Series in the Cross Fell Inlier. *Proc. Yorks. geol. Soc.*, **32**, 185–228.

—— 1962. The trilobites of the Caradoc Series in the Cross Fell Inlier of northern England. *Bull. Br. Mus. nat. Hist.* (Geol.), **7**, (3), 67–134, pls. 6–18.

—— 1963. The Ordovician trilobite faunas of south Shropshire. III. *Ibid.*, **7**, (8), 213–54, pls. 37–46.

—— 1963a. The Stile End Beds and Drygill Shales (Ordovician) in the East and North of the English Lake District. *Ibid.*, **9**, (3), 47–65, pls. 1–5.

—— 1964. The geology of the Ordovician and adjacent strata in the southern Caradoc district of Shropshire. *Ibid.*, **9**, (7), 257–96, pls. 1, 2.

—— 1966. The Lower Ordovician stratigraphy and trilobites of the Landeyran Valley and the neighbouring district of the Montagne Noire, south-western France. *Ibid.*, **12**, (6), 245–353, pls. 1–21.

—— 1971. The trilobites of the Chair of Kildare Limestone (Upper Ordovician) in eastern Ireland. *Palaeontogr. Soc.* [*Monogr.*]: (1), 1–60, pls. 1–25.

EMMRICH, H. 1839. *De Trilobitis. Dissertatio Petrefactologica.* Berolini.

ERBEN, H. K. 1951. Beitrag zur Gliederung der Gattung *Proetus* STEIN. 1831 (Trilobitae). *Neues Jb. Geol. Paläont.*, *Abh.*, **94**, (1), 5–48, pls. 2–4.

—— 1966. Über die Tropidocoryphinae (Tril.) Lfg. 1. *Ibid.*, **125**, 170–211, pls. 19–21.

ESMARK, H. M. T. 1833. Om nogle nye Arter af Trilobiter. *Mag. f. Naturvid.*, **11**, 268–70, pl. 7.

ETHERIDGE, R. & MITCHELL, J. 1892. The Silurian trilobites of New South Wales, with reference to those of other parts of Australia. Part 1. *Proc. Linn. Soc. New South Wales*, **6**, 311–20, pl. 25.

GOLDFUSS, A. 1843. Systematische Übersicht der Trilobiten und Beschreibung einiger neuen Arten derselben. *Neues Jb. Mineral.* for 1843, 537–67, pls. 4–6.

HAAS, W. 1969. Lower Devonian trilobites from central Nevada and northern Mexico. *J. Paleont.*, **43**, 641–59, pls. 81–84.

HAHN, G. & HAHN, R. 1967. Zur Phylogenie der Proetidae (Trilobita) des Karbons und Perms. *Zool. Beitr. N.F.*, **13**, 303–49. Berlin.

HARPER, J. C. 1952. The Ordovician rocks between Collon (Co. Louth) and Grangegeeth (Co. Meath). *Sci. Proc. Roy. Dublin Soc.*, (N.S.), **26**, 85–112, pls. 5–7.

—— 1956. The Ordovician succession near Llanystwmdwy, Caernarvonshire. *Lpool Manchr geol. J.*, **1**, (4), 385–93.

HAWLE, I. & CORDA, A. J. C. 1847. *Prodrom einer Monographie der böhmischen Trilobiten.* 176 pp., 7 pls. Prague.

HEDE, J. E. 1915. Skånes Colonusskiffer. *Acta Univ. lund.* N.F., Avd. 2, **11**, (6), 1–65, pls. 1–4.

HEDSTRÖM, H. 1923. Contributions to the fossil fauna of Gotland. I. *Sver. geol. Unders. Afh.*, Ser. C, **316**, 1–24, pls. 1–5.

HESSLER, R. R. 1962. The Lower Mississippian genus *Proetides* (Tril.). *J. Paleont.*, **36**, (4), 811–6, pl. 119.

—— 1963. Lower Mississippian trilobites of the family Proetidae in the United States part 1. *Ibid.*, **37**, (3), 543–63, pls. 59–62.

HORNÝ, R. & BASTL, F. 1970. *Type specimens of fossils in the National Museum, Prague. Volume 1. Trilobita.* 354 pp., 20 pls. National Museum, Prague.

—— PRANTL, F. & VANĚK, J. 1958. [On the limit between the Wenlock and the Ludlow in the Barrandian.] *Sbor. úst. Úst. geol.*, *Odd. Palaeont.*, **24**, for 1957, 217–78, pls. 29–37. [In Czech, with English summary].

HUGHES, C. P. 1969. The Ordovician trilobite faunas of the Builth-Llandrindod Inlier, central Wales. Part 1. *Bull. Br. Mus. nat. Hist.* (Geol.), **18**, (3), 41–103, pls. 1–14.

HUPÉ, P. 1953. *Classe de trilobites.* In PIVETEAU, J. (Editor). *Traité de paléontologie*, **3**, 44–246, 140 text-figs. Paris.

HUXLEY, T. H., NEWTON, E. T. & ETHERIDGE, R. 1878. *A catalogue of the Cambrian and Silurian fossils in the Museum of Practical Geology.* iii+144 pp. London.

INGHAM, J. K. 1966. The Ordovician rocks in the Cautley and Dent districts of Westmorland and Yorkshire. *Proc. Yorks. geol. Soc.*, **35**, 455–505.

—— 1970. A monograph of the Upper Ordovician trilobites from the Cautley and Dent districts of Westmorland and Yorkshire. *Palaeontogr. Soc.* [*Monogr.*]: (1), 1–58, pls. 1–9.

KEGEL, W. 1927. Über obersilurische Trilobiten aus dem Harz und dem Rheinischen Schiefergebirge. *Jb. preuss. geol. Landesanst.*, **48**, 616–47, pls. 31–2.

KIELAN, Z. 1960. Upper Ordovician trilobites from Poland and some related forms from Bohemia and Scandinavia. *Palaeont. pol.*, **11**, I–VI, 1–198, pls. 1–36. (Dated 1959).

KING, W. B. R. 1932. A fossiliferous limestone associated with Ingletonian beds at Horton-in-Ribblesdale, Yorkshire. *Q. Jl geol. Soc. Lond.*, **88**, 101–11.

—— & WILLIAMS, A. 1948. On the lower part of the Ashgillian Series in the north of England. *Geol. Mag.*, **85**, 205–12, pl. 16.

LINDSTRÖM, G. 1885. Förteckning på Gotlands siluriska crustacéer. *Öfvers K. Vetensk Akad. Forh.*, **6**, 37–100, pls. 12–16.
—— 1901. Researches on the visual organs of the trilobites. *K. svenska Vetensk.-Akad. Handl.*, **34**, 1–89, pls. 1–6.
LOVÉN, S. 1845. Svenska Trilobiter. *Öfvers K. Vetensk Akad. Förh.*, **2**, 46–56, pl. 1, 104–11, pl. 2.
LU, Y.-H. 1962. [Restudy of Grabau's types of three Silurian trilobites from Hupeh.] *Acta Pal. Sinica*, **10**, (2), 158–74, pl. 1. [In Chinese, with English summary].
MACGREGOR, A. R. 1963. Upper Llandeilo trilobites from the Berwyn Hills, North Wales. *Palaeontology*, **5**, 790–816, pls. 116–118.
MAKSIMOVA, Z. A. 1970. [*Silurian trilobites of Vajgač Island*]. In: [*Silurian stratigraphy and fauna of Vajgač Island*]. *Nauchno-issled. Inst. Geol. Arktiki*, 195–209, pls. 1, 2. [In Russian].
MARR, J. E. 1916. The Ashgillian succession in the tract to the west of Coniston Lake. *Q. Jl geol. Soc. Lond.*, **71**, 189–204.
—— & NICHOLSON, H. A. 1888. The Stockdale Shales. *Q. Jl geol. Soc. Lond.*, **44**, 654–732, pl. 16.
MARTINSSON, A. 1968. Towards a new style in palaeontological publishing. *Lethaia*, **1**, I, II.
MCCOY, F. 1846. *A synopsis of the Silurian fossils of Ireland.* 72pp., 5 pls. Dublin.
—— 1854. *Contributions to British palaeontology or first descriptions of three hundred and sixty species and several genera of fossil Radiata, Articulata, Mollusca, and Pisces from the Tertiary, Cretaceous, Oolitic, and Palaeozoic strata of Great Britain.* 272 pp., frontispiece, many figs. (not numbered). Cambridge. (Republished from *Ann. Mag. nat. Hist.*).
MITCHELL, G. H., POCOCK, R. W. & TAYLOR, J. H. 1961. Geology of the country around Droitwich, Abberley and Kidderminster. *Mem. geol. Surv. U.K.*, x+137 pp., 3 pls.
MOORE, R. C. (Editor). 1959. *Treatise on Invertebrate Paleontology, Part O, Arthropoda 1.* xix+560 pp., 415 figs. Geol. Soc. Amer. & Univ. Kansas Press (Lawrence).
MORRIS, J. 1854. *A catalogue of British fossils: comprising the genera and species hitherto described; with references to their geological distribution and to the localities in which they have been found.* (2nd edition). viii+372 pp. London.
MURCHISON, R. I. 1839. *The Silurian System, founded on geological researches in the counties of Salop, Hereford, Radnor, Montgomery, Caermarthen, Brecon, Pembroke, Monmouth, Gloucester, Worcester and Stafford; with descriptions of the coalfields and overlying formations.* xxxii+768 pp., 37 pls. London.
—— 1854. *Siluria. The history of the oldest known rocks containing organic remains, with a brief description of the distribution of gold over the earth.* xvi+523 pp., 37 pls.
—— 1859. *Ibid.* 3rd edition. xx+592 pp., 41 pls.
—— 1867. *Ibid.* 4th edition. xviii+566 pp., 41 pls.
—— 1872. *Ibid.* 5th edition. xviii+566 pp., 41 pls.
NICHOLSON, H. A. & ETHERIDGE, R. 1879. *A monograph of the Silurian fossils of Girvan in Ayrshire with special reference to those contained in the 'Gray Collection'.* Vol. 1: (2), i–vi, 137–236, pls. 10–15. Edinburgh & London.
NIESZKOWSKI, J. 1857. Versuch einer Monographie der in den silurischen Schichten der Ostseeprovinzen vorkommenden Trilobiten. *Archiv Naturk. Liv-Ehst-Kurlands*, (1), **1**, 517–626, pls. 1–3.
NOVÁK, O. 1890. Vergleichende Studien an einigen Trilobiten aus dem Hercyn von Bicken, Wildungen, Greifenstein und Böhmen. *Paläont. Abh., n.F.*, **1**, (3), 1–46, pls. 19–23. Jena.
OEHLERT, M. D. 1886. Étude sur quelques trilobites du groupe de Proetidae. *Bull. Soc. Et. sci. Angers, n.S.*, **15**, 121–43, pls. 1, 2. Angers.
ÖPIK, A. A. 1937. Trilobiten aus Estland. *Acta Comment. Univ. Tartu*, (A), **32**, (3), 1–163, pls. 1–26 (*Pub. Geol. Inst. Univ. Tartu*, no. 52).
ORMISTON, A. R. 1967. Lower and Middle Devonian trilobites of the Canadian Arctic Islands. *Bull. Geol. Surv. Canada*, **153**, 1–148, 17 pls.
—— 1971. Lower Devonian trilobites from the Michelle Formation, Yukon territory. *Ibid.*, **192**, 27–44, pls. 3–4.
—— 1971a. Silicified specimens of the Gedinnian trilobite *Warburgella rugulosa canadensis* Ormiston, from the Northwest Territories Canada. *Paläont. Z.*, **45**, 173–80, pls. 19–21.
OSMÓLSKA, H. 1970. Revision of non-cyrtosymbolinid trilobites from the Tournaisian-Namurian of Eurasia. *Palaeont. pol.*, **23**, 1–165, pls. 1–22.
OWENS, R. M. 1970. The Middle Ordovician of the Oslo Region, Norway. 23. The trilobite family Proetidae. *Norsk geol. Tidsskr.*, **50**, 309–32.
—— 1973. Ordovician Proetidae (Trilobita) from Scandinavia. *Ibid.*, **53**, 117–81.
PILLET, J. 1969. La classification des Proetidae (Trilobites) [1]. *Bull. Soc. Etud. scient. Anjou, N.S.*, **7**, 53–85, pls. 1–6.
POULSEN, C. 1934. The Silurian faunas of North Greenland, 1. The fauna of the Cape Schuchert Formation. *Meddr Grønland*, **72**, (1), 1–46, pls. 1–3.
PŘIBYL, A. 1946. O několika nových trilobitových rodech z českého siluru a devonu. *Příroda, Brno*, **38**, (5–6), 7 pp., 11 text-figs.
—— 1946a. Příspevek k poznáni českých Proetidů. *Rozpr. II tř. české Akad.*, **55**, (10), 1–37. Prague. (Notes on the recognition of the Bohemian Proetidae. *Bull. int. Acad. tchéque Sci.* 1945, **46**, 1–41, pls. 1–4. Prague).
—— 1947. *Proetus (Eremiproetus?) Reedi* n. nom. nový druh trilobitu ze Skotského ordoviku. *Příroda, Brno*, **39**, (1), 4 pp., 6 text-figs.

Přibyl, A. 1949. O několika nových nebo málo známých trilobitech z českého devonu. (On several new or little known trilobites of the Devonian of Bohemia). *Věstn. státn. geol. Ústav. ČSR.*, **24**, 293–330, pls. 1, 2. Prague. [In Czech, with English summary].

—— 1953. Seznam českých trilobitových rodů (Index of trilobite genera in Bohemia). *Knihovna Ústř. úst. geol.*, Svazek, **25**, 80 pp. [Czech, English and Russian texts].

—— 1964. Neue Trilobiten (Proetidae) aus dem böhmischen Devon. *Spis. bulg. geol. Druzh.*, **25**, (1), 23–51, pls. 1–3. Sofia.

—— 1965. Proetidní triloboti z nových sběrů v českém siluru a devonu. (Proetidae aus neueren Aufsammlungen im böhmischen Silur und Devon (Trilobitae) I). *Čas. národ. Mus., Odd. přírod.*, **134**, (2), 91–8, pls. 7, 8. Prague. [In German, with Czech summary].

—— 1970. O několika českých a asijských zástupcích proetidních trilobitů. (Über einige böhmische und asiatische Vertreter von Proetiden (Trilobita)). *Čas. Miner. Geol.*, **15**, 101–11, 1 pl. [In German, with Czech summary].

—— & Vaněk, J. 1962. Trilobitová fauna českého svrchního siluru (budňanu a lochkovu) a její biostratigrafický význam. (Die Trilobiten-Fauna aus dem böhmischen Obersilur (Budnanium und Lochkovium) und ihre biostratigraphische Bedeutung). *Sborn. národ. Muzea Praze*, **18**, B, 25–46, pls. 1–4. [In German, with Czech summary].

Reed, F. R. C. 1901. Woodwardian Museum notes: Salter's undescribed species. 2. *Geol. Mag.*, (4), **8**, 5–14, pl. 1.

—— 1904. The Lower Palaeozoic trilobites of the Girvan district, Ayrshire. *Palaeontogr. Soc.* [*Monogr.*]: (2), 49–96, pls. 7–13.

—— 1914. The Lower Palaeozoic trilobites of Girvan. Supplement. *Ibid.*, 56 pp., 8 pls.

—— 1916. Palaeontological appendix (*in* Gardiner, C. I. The Silurian Inlier of Usk). *Proc. Cotteswold Nat. Fld Club*, **19**, 160–72, pls. 7, 8.

—— 1931. The Lower Palaeozoic trilobites of Girvan. Supplement No. 2. *Palaeontogr. Soc.* [*Monogr.*], 30 pp.

—— 1935. The Lower Palaeozoic trilobites of Girvan. Supplement No. 3. *Ibid.*, 64 pp., 4 pls.

—— 1940. New Ordovician fossils from Girvan, Ayrshire. *Ann. Mag. nat. Hist.*, (11), **6**, 154–60, pl. 8.

—— 1941. A new genus of trilobites and other fossils from Girvan. *Geol. Mag.*, **78**, 268–78, pl. 5.

Richter, Reinh. 1863. Aus dem thüringischen Schiefergebirge. *Z. dt. geol. Ges.*, **15**, 659–76, pls. 18, 19.

Richter, Rud. 1912. Beitrage zur Kenntnis devonischer Trilobiten. I. Die Gattung *Dechenella* und einige verwandte Formen. *Abh. Senck. Nat. Gesell.*, **31**, 239–340, pls. 18–21.

—— & Richter, E. 1918. Neue Proetus-Arten aus dem Eifler Mittel-Devon. *Zentbl. Miner.*, (1918), 64–70.

—— —— 1919. Der Proetiden Zwieg *Astycoryphe-Tropidocoryphe-Pteroparia. Senckenbergiana*, **1**, 1–17, 25–51.

—— —— 1923. Der Genotyp von *Proetus* Stein., 1831. *Senckenbergiana*, **5**, 240.

—— —— 1954. Die Trilobiten des Ebbe-Sattels und zu vergleichende Arten (Ordovicium, Gotlandium/Devon). *Abh. Senck. Nat. Gesell.*, **488**, 1–176, 6 pls.

—— —— 1956. Annular-Teilung bei Trilobiten am Beispiel besonders von *Proetus* (*Pr.*) *cuvieri* und *prox. Senckenb. leth.*, **37**, 343–81, 6 pls.

Rickards, R. B. 1965. Two new genera of Silurian phacopid trilobites. *Palaeontology*, **7**, 541–51, pls. 84–5.

—— 1967. The Wenlock and Ludlow succession in the Howgill Fells (north-west Yorkshire and Westmorland). *Q. Jl geol. Soc. Lond.*, **123**, 215–51.

Romano, M. & Diggens, J. N. 1969. Longvillian shelly faunas from the Dolwyddelan area, north Wales. *Geol. Mag.*, **106**, 603–6.

Salter, J. W. 1848. *In* Phillips, J. & Salter, J. W. Palaeontological appendix to Professor John Phillips' memoir on the Malvern Hills, compared with the Palaeozoic districts of Abberley, etc. *Mem. geol. Surv. U.K.*, **2**, (1), viii–xiv+331–86, pls. 4–30.

—— 1864. A monograph of the British trilobites from the Cambrian, Silurian and Devonian formations. *Palaeontogr. Soc.* [*Monogr.*]: (1), 1–80, pls. 1–6.

—— 1873. *A Catalogue of the collections of Cambrian and Silurian fossils contained in the Geological Museum of the University of Cambridge.* ix–xlviii, 204 pp. Cambridge.

Sandberger, G. & Sandberger, F. 1850. *Die Versteinerungen des Rheinischen Schichtensystems in Nassau.* 564 pp., 39 pls. Wiesbaden.

Schmidt, F. 1894. Revision der Ostbaltischen Silurischen Trilobiten, Abt. 4. *Zap. imp. Akad. Nauk* [=*Mém. Acad. imp. Sci. St.–Pétersb.*], (7), **42**, (5), 1–94, pls. 1–6.

Sdzuy, K. 1955. Die Fauna der Leimitz-Schiefer (Tremadoc). *Abh. Senck. Nat. Gesell.*, **492**, 1–74, pls. 1–8.

Selwood, E. B. 1965. Dechenellid trilobites from the British Middle Devonian. *Bull. Br. Mus. nat. His.* (Geol.), **10**, (9), 317–33, 1 pl.

Sjöberg, S. 1918. Beschreibung einer neuen Trilobitenart. *Geol. Fören. Stockh. Förh.*, **40**, 457–60, pl. 7.

Smyčka, F. 1895. Devonští triloboti u Čelechovic na Moravě. *Rozpr. české Akad.*, **4**, 1–15, pl. 1.

Steininger, J. 1831. Observations sur les fossiles du Calcaire intermédiare de l'Eifel. *Mém. Soc. géol. Fr.*, **1** (1), 331–71, pls. 21–23.

Strahan, A., Cantrill, T. C., Dixon, E. E. L. & Thomas, H. H. 1907. The geology of the South Wales Coalfield. Part VII. The country around Ammanford. *Mem. geol. Surv. U.K.*, i–viii, 1–246.

—— —— —— —— 1909. The geology of the South Wales Coalfield. Part X. The country around Carmarthen. *Ibid.*, i–viii, 1–177.

—— —— —— —— & Jones, O. T. 1914. The geology of the South Wales Coalfield. Part XI. The country around Haverfordwest. *Ibid.*, i–viii, 1–262.

Stubblefield, C. J. 1938. The types and figured specimens in Phillips' and Salter's Palaeontological Appendix to Phillips' Memoir on "The Malvern Hills compared with the Palaeozoic Districts of Abberley etc." (*Mem. geol. Surv. U.K.*, **2**, (1), June 1848). *Summ. Progr. geol. Surv. U.K.* (for 1936), (2), 27–51.

Stumm, E. C. 1953. Trilobites of the Devonian Traverse Group of Michigan. *Contr. Mus. Paleont. Univ. Mich.*, **10**, (6), 101–44, pls. 1–12.

Temple, J. T. 1969. Lower Llandovery (Silurian) trilobites from Keisley, Westmorland. *Bull. Br. Mus. nat. Hist.* (Geol.), **18**, (6), 197–230, 6 pls.

—— 1970. The Lower Llandovery brachiopods and trilobites from Ffridd Mathrafal, near Meifod, Montgomeryshire. *Palaeontogr. Soc.* [*Monogr.*], 76 pp., 19 pls.

Törnquist, S. L. 1884. Undersökningar öfver Siljansområdets trilobitfauna. *Sver. geol. Unders. Afh.*, Ser. C, **66**, 101 pp., 3 pls.

Tripp, R. P. 1954. Caradocian trilobites from mudstones at Craighead Quarry, near Girvan, Ayrshire. *Trans. R. Soc. Edinb.*, **62**, (3), 655–93, pls. 1–4.

—— 1962. Trilobites from the "*confinis*" flags (Ordovician) of the Girvan district, Ayrshire. *Ibid.*, **65**, (1), 1–40, pls. 1–4.

—— 1967. Trilobites from the Upper Stinchar Limestone (Ordovician) of the Girvan district, Ayrshire. *Ibid.*, **67**, (3), 43–93, pls. 1–6.

Wade, A. 1911. The Llandovery and associated rocks of north-eastern Montgomeryshire. *Q. Jl geol. Soc. Lond.*, **67**, 415–59, pls. 33–36.

Warburg, E. 1925. The trilobites of the Leptaena Limestone in Dalarne. *Bull. geol. Instn Univ. Upsala*, **17**, (4), i–vi, 1–446, pls. 1–11.

Whittard, W. F. 1928. The stratigraphy of the Valentian rocks of Shropshire. The Main Outcrop. *Q. Jl geol. Soc. Lond.*, **83**, [for 1927], 737–59, pls. 56–57.

—— 1938. The Upper Valentian trilobite fauna of Shropshire. *Ann. Mag. nat. Hist.*, (11), **1**, 85–140, pls. 2–4.

—— 1961, 1966. The Ordovician trilobites of the Shelve Inlier, West Shropshire. *Palaeontogr. Soc.* [*Monogr.*]: (5), 1961, 163–96, pls. 21–25; (8), 1966, 265–306, pls. 46–50.

Whittington, H. B. 1950. A monograph of the British trilobites of the family Harpidae. *Palaeontogr. Soc.* [*Monogr.*], ii +55 pp., 7 pls.

—— 1960. *Cordania* and other trilobites from the Lower and Middle Devonian. *J. Paleont.*, **34**, (3), 405–20, pls. 51–54.

—— 1963. Middle Ordovician trilobites from Lower Head, western Newfoundland. *Bull. Mus. comp. Zool. Harv.*, **129**, 1–118, pls. 1–36.

—— 1966. Phylogeny and distribution of Ordovician trilobites. *J. Paleont.*, **40**, (3), 696–737.

—— 1966a. A monograph of the Ordovician trilobites of the Bala area, Merioneth. *Palaeontogr. Soc.* [*Monogr.*]: (3), 63–92, pls. 19–28.

—— & Campbell, K. S. W. 1967. Silicified Silurian trilobites from Maine. *Bull. Mus. comp. Zool. Harv.*, **135**, 447–82, pls. 1–19.

Williams, A., Strachan, I., Bassett, D. A., Dean, W. T., Ingham, J. K. *et al.* 1972. A correlation of Ordovician rocks in the British Isles. *Geol. Soc. Lond. Spec. Rep.*, no. 3, 1–74 (undated).

Woods, H. 1891. *Catalogue of the type fossils in the Woodwardian Museum, Cambridge.* xiv+180 pp. Cambridge.

Woodward, H. 1877. *A catalogue of British fossil Crustacea, with their synonyms and the range in time of each genus and order.* xii+147 pp. London.

Yolkin, E. A. 1966. [A new genus and new species of Lower Devonian and Eifelian Dechenellidae (Trilobites)]. *Akad. Nauk. SSSR. sib. otd. Inst. Geol. Geofiz.*, 1966, (2), 25–30. [In Russian].

—— 1968. [Trilobites (Dechenellidae) and stratigraphy of the Lower and Middle Devonian of southern West Siberia]. *Ibid.*, 1968, 1–155, pls. 1–13. [In Russian].

Ziegler, A. M., Cocks, L. R. M. & McKerrow, W. S. 1968. The Llandovery transgression of the Welsh Borderland. *Palaeontology*, **11**, 736–82.

Robert M. Owens, Ph.D.,
Department of Geology,
National Museum of Wales,
Cardiff,
CF1 3NP.

INDEX

Numbers in bold type indicate pages on which a description is given or commences; numbers in italics indicate pages where a text-figure occurs; Plate and figure numbers are also given, as are the names of all stratigraphical units and selected localities. Names of taxa included in square brackets are considered invalid.

EXPLANATION OF PLATES

Before photographing, the specimens were given a light coating of ammonium chloride sublimate, some having previously received a thin coat of dilute black 'opaque' to produce more even contrast. Three small ontogenetic stages (Pl. 7, figs. 13, 16, 17) were photographed with a scanning electron microscope, and were coated with a thin film of aluminium.

For the illustration of some specimens (e.g. Pl. 1, figs. 1, 11a) stereo pairs are used. The pairs of photographs are so spaced that the stereo-effect can be achieved either with a stereo viewer or with unaided eyes.

Abbreviations of repositories of figured specimens:

BM	British Museum (Natural History), London
BU	Birmingham University, Department of Geology
GSM	Geological Survey and Museum, Institute of Geological Sciences, London
HM	Hunterian Museum, Glasgow
HUD	Hull University, Department of Geology, Ingham Collection
HUR	Hull University, Department of Geology, Rickards Collection
ICMM	Imperial College, London, Department of Geology, Murchison Museum
LCM	Leicester City Museum and Art Gallery
LPI	Palaeontological Institute, University of Lund, Sweden
LRU	Leicester University, Department of Geology
LU	Liverpool University, Department of Geology
NMI	National Museum of Ireland, Dublin
NMP	National Museum, Prague, Czechoslovakia
NMW	National Museum of Wales, Cardiff
OUM	Oxford University Museum
RM	Naturhistoriska Riksmuseet, Stockholm, Sweden
SM	Sedgwick Museum, Cambridge
SMF	Senckenberg Museum, Frankfurt-am-Main, Germany
TCD	Trinity College, Dublin, Department of Geology
UM	Palaeontological Institute, University of Uppsala, Sweden

PLATE 1

Plate 1

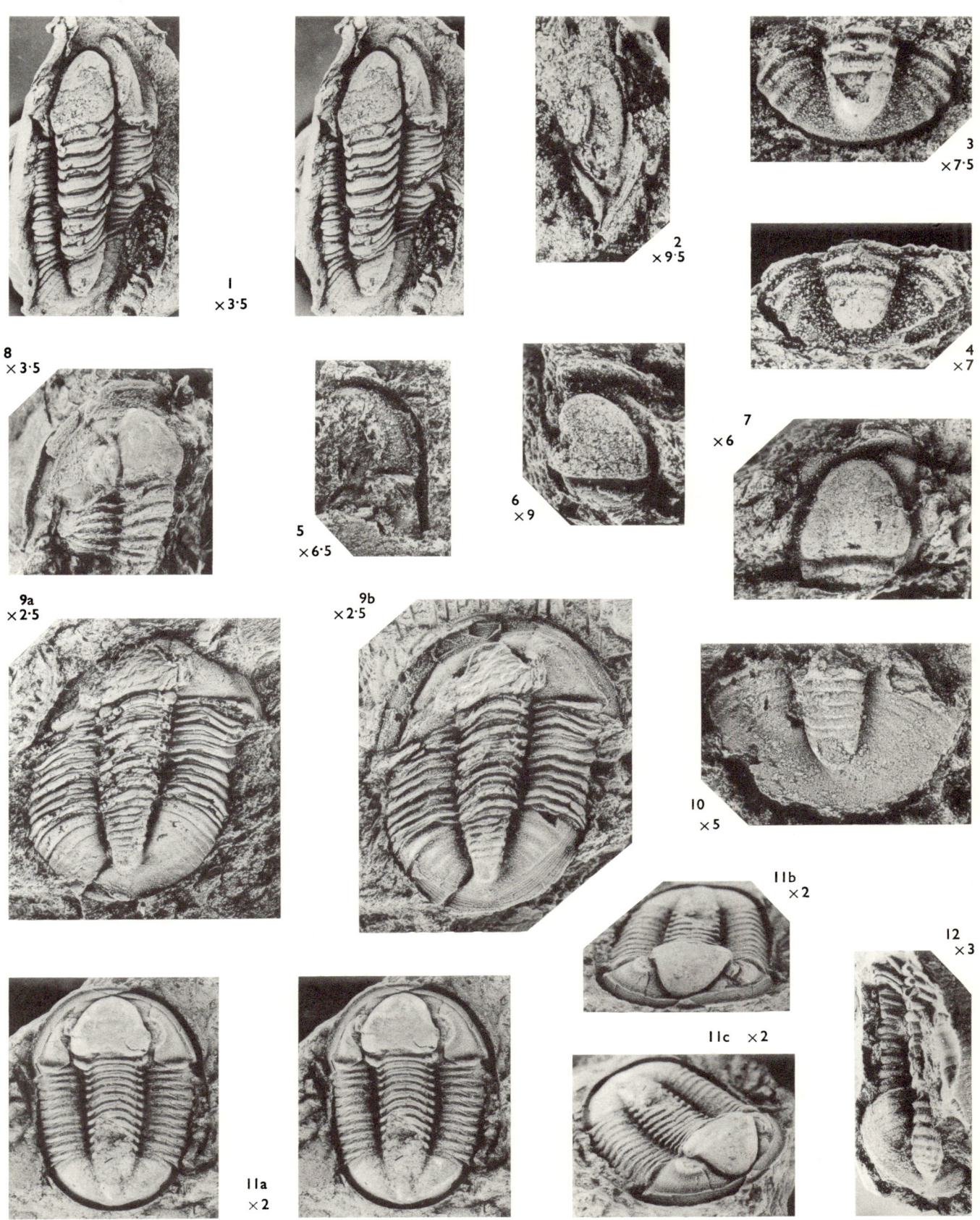

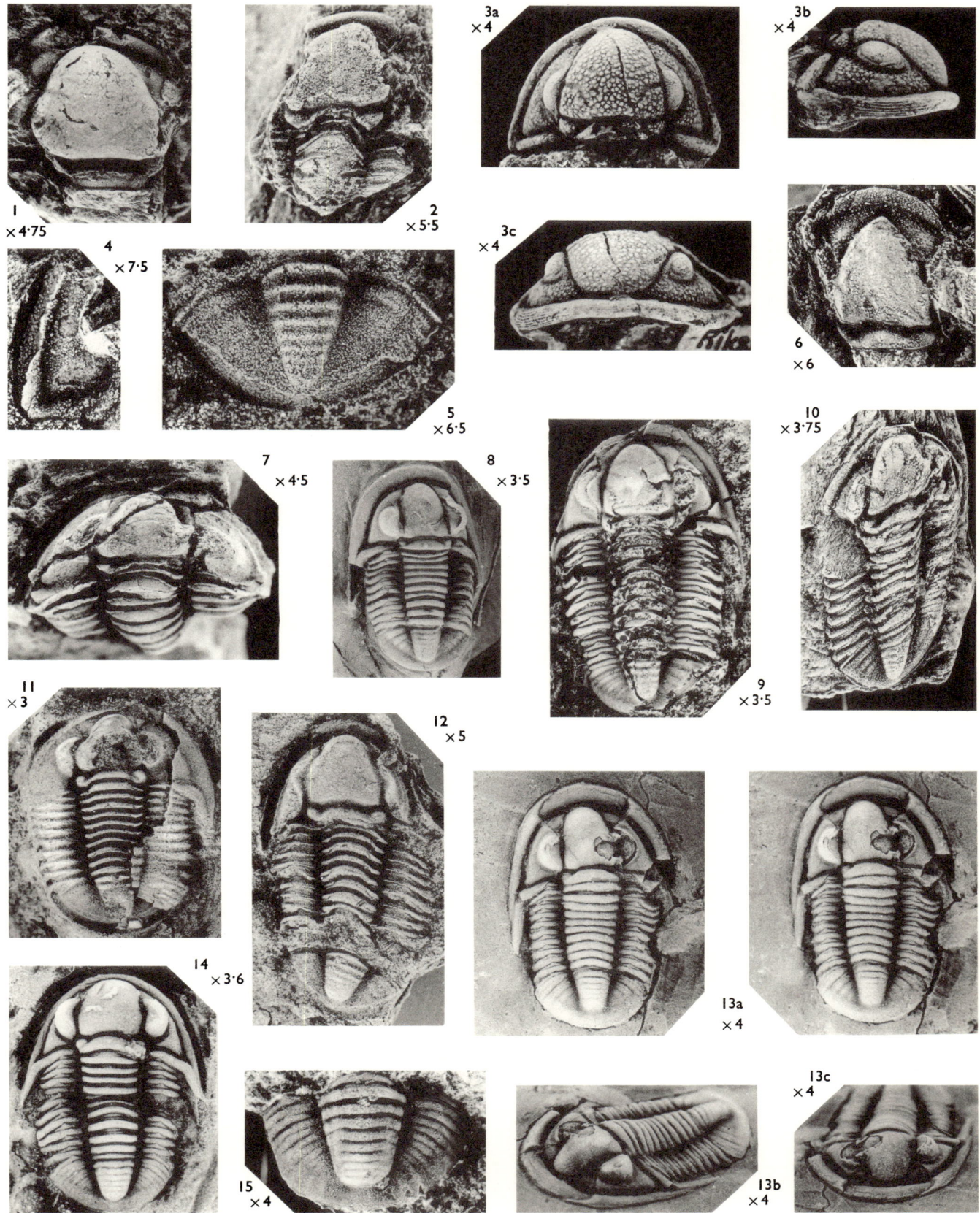

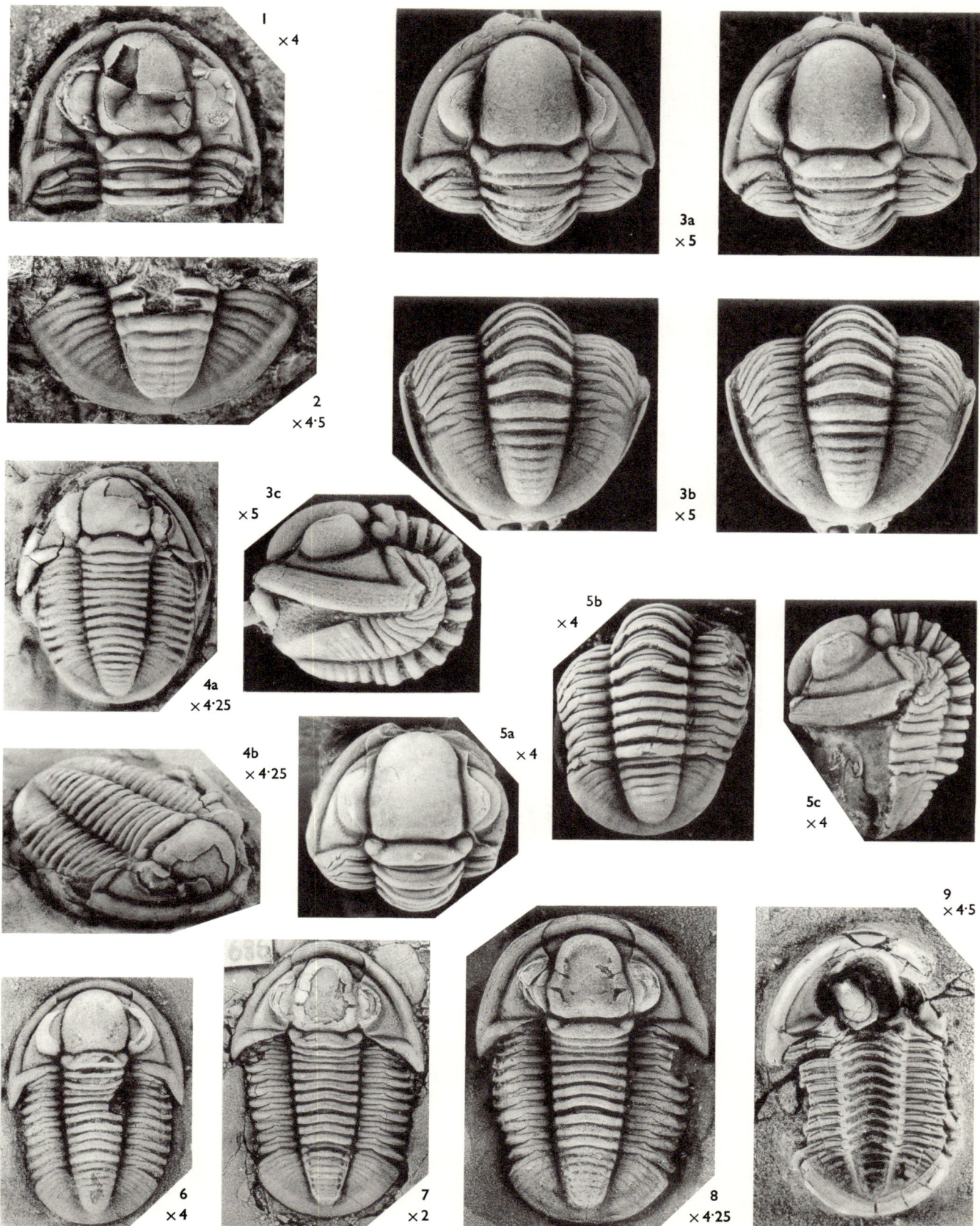

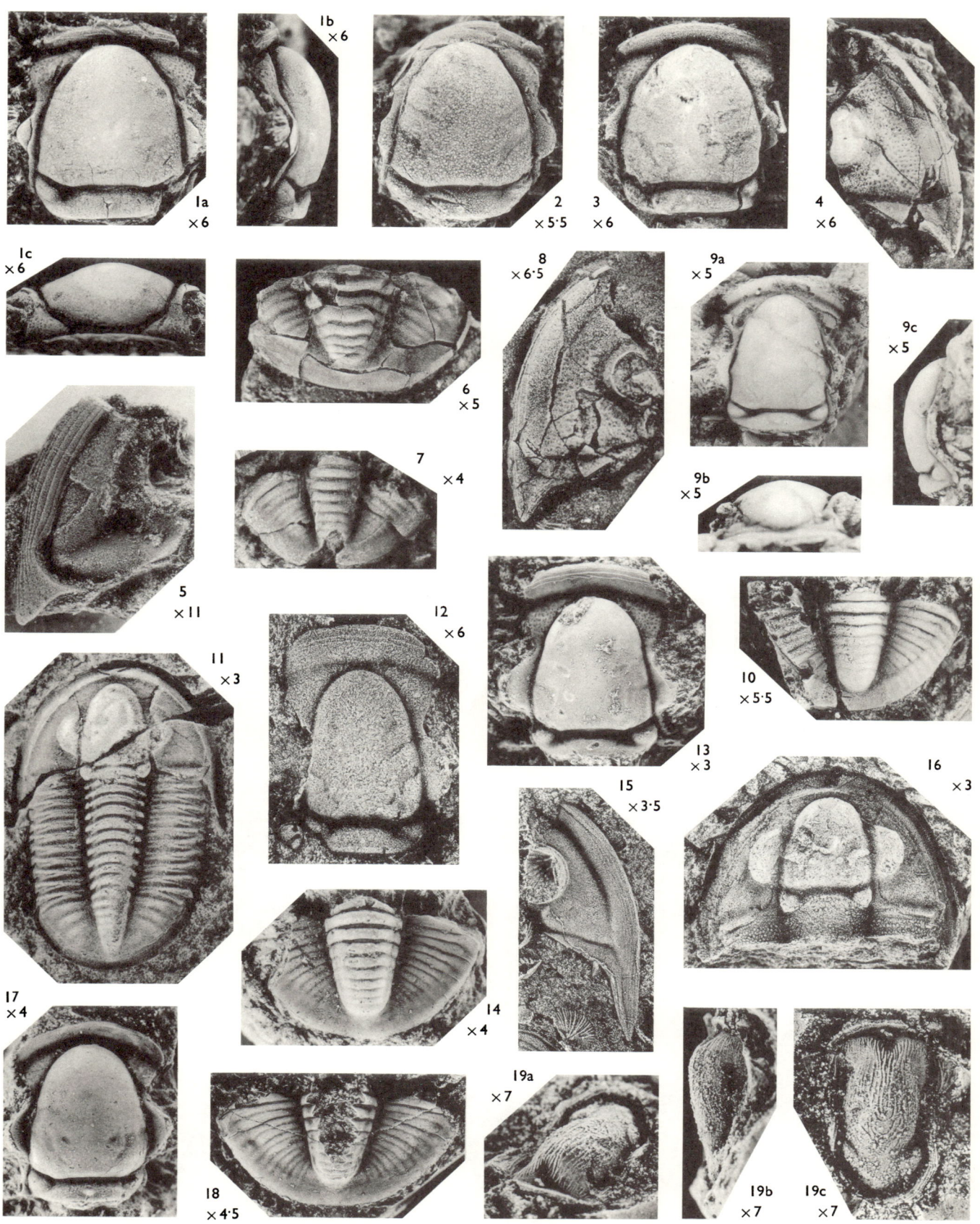

PLATE 5

Plate 5

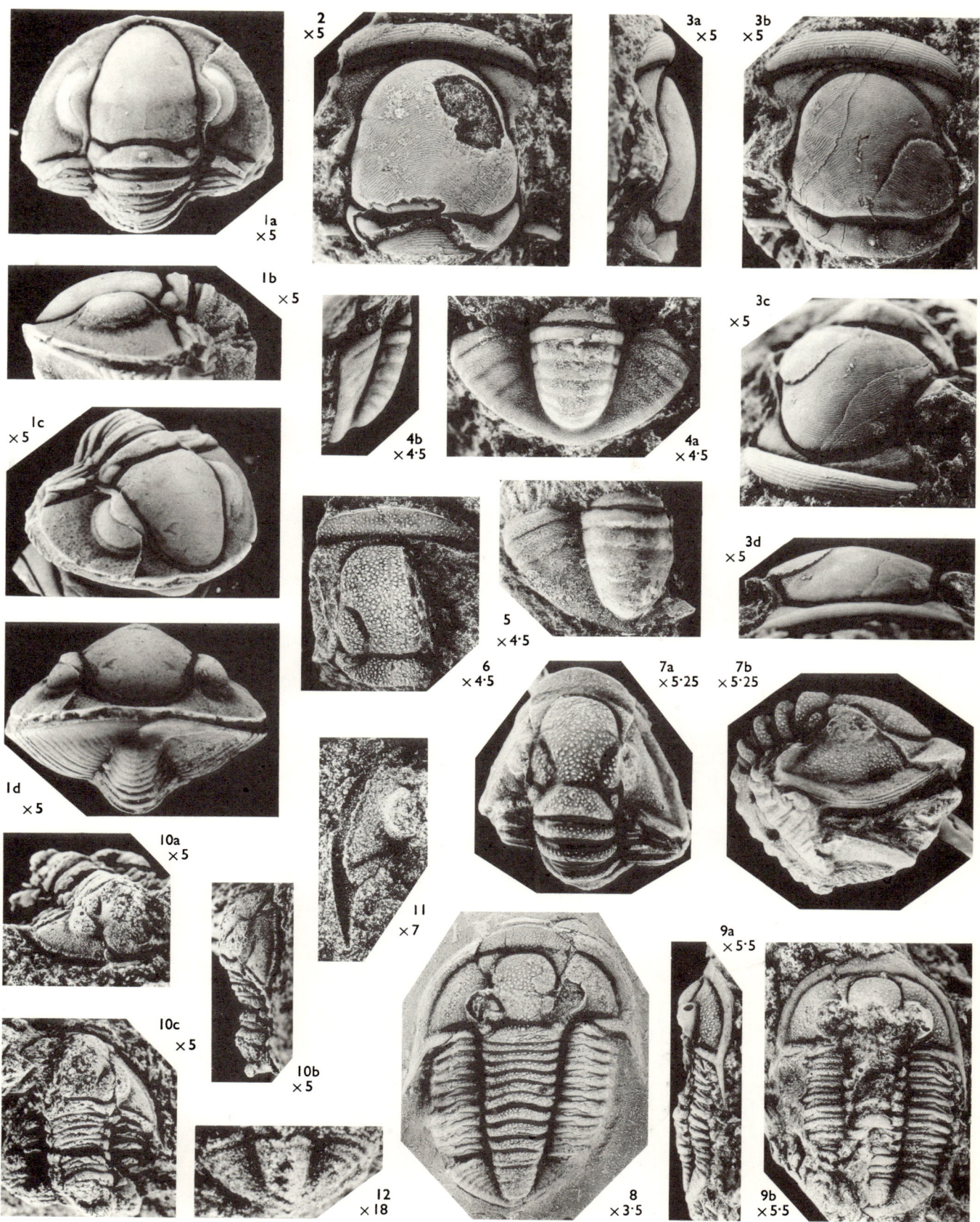

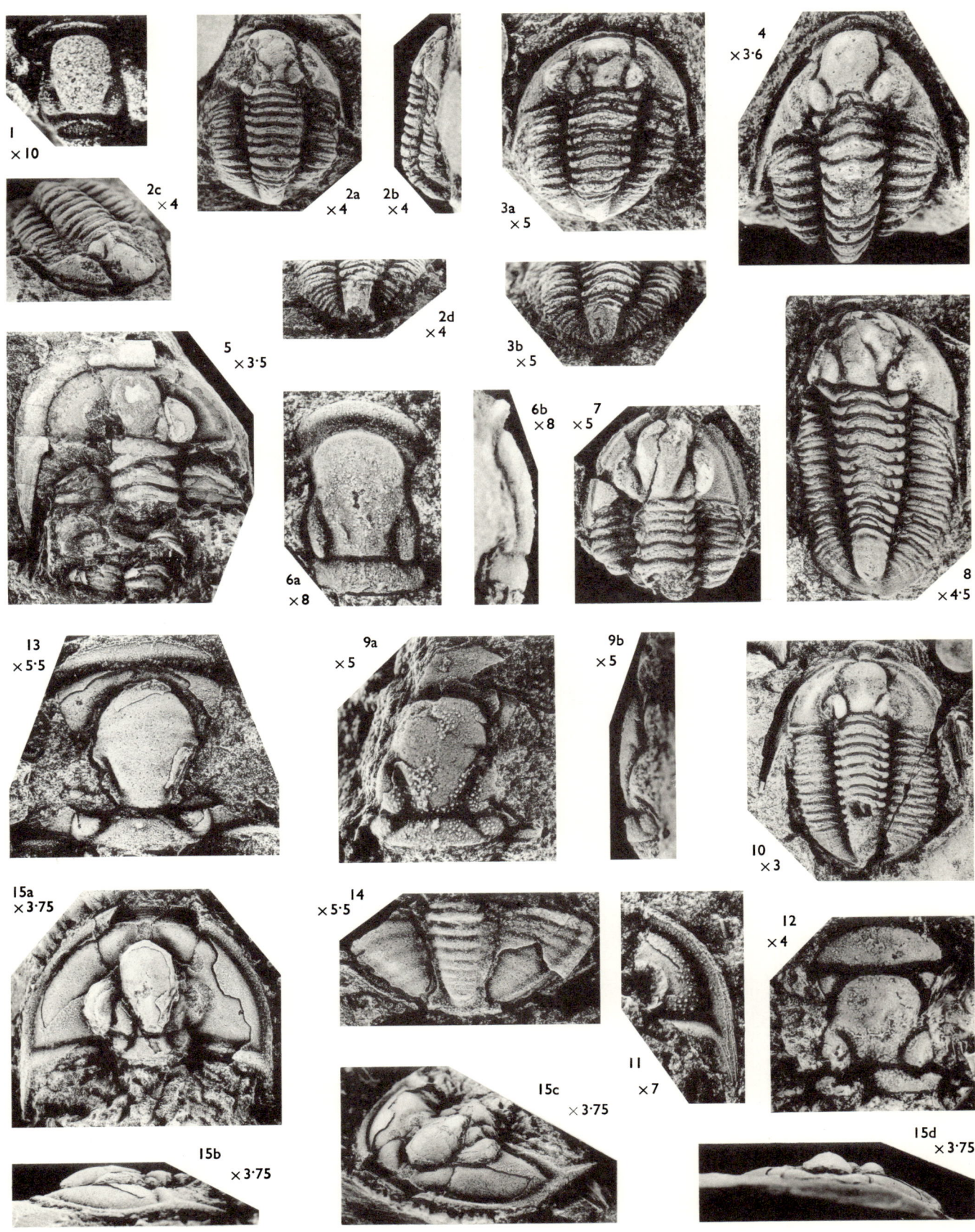

Plate 7

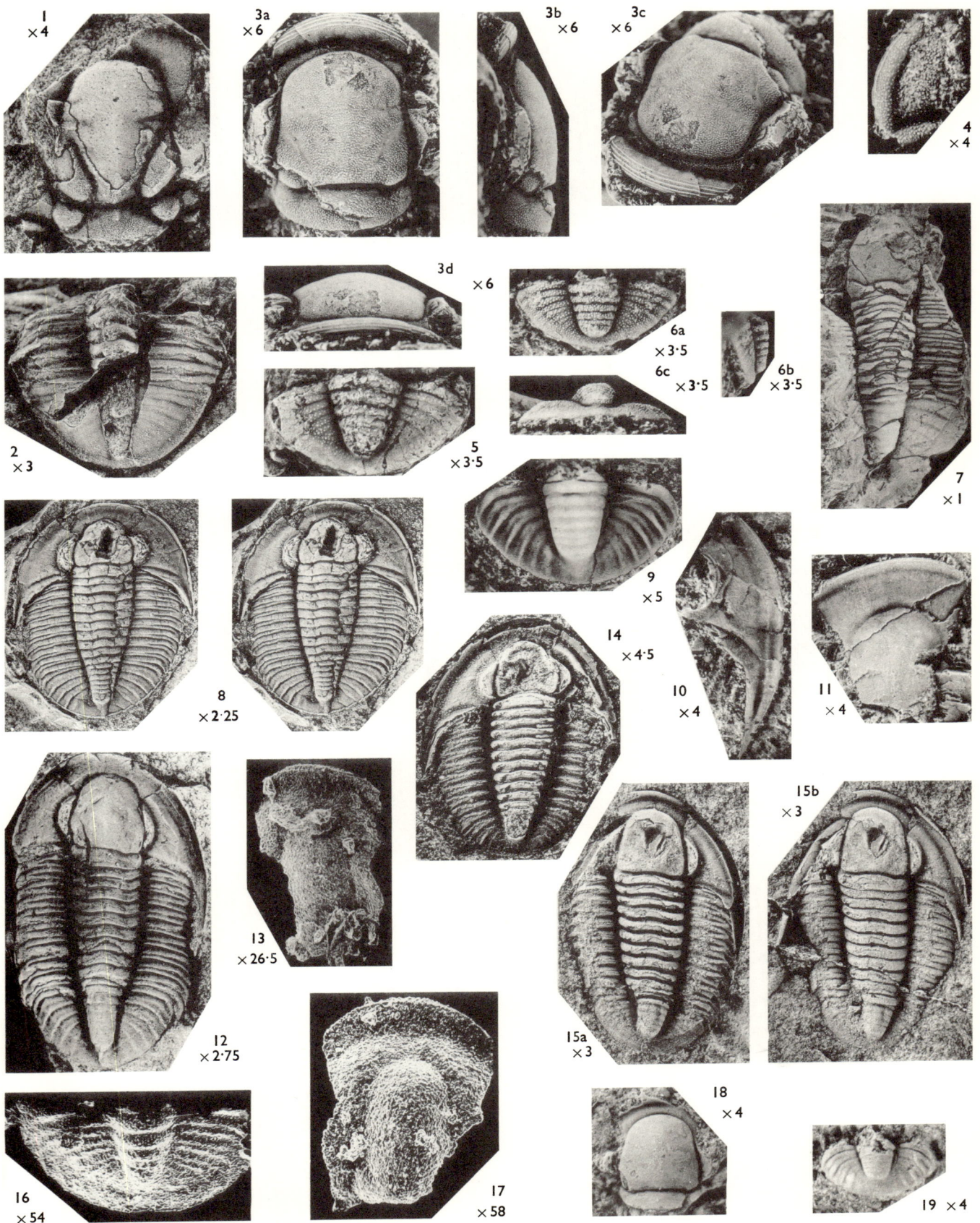

PLATE 8

Decoroproetus calvus (Whittard, 1961)

1a, b. Holotype cranidium, internal mould, dorsal view. × 3·5. Near base of Whittery Shales (Caradoc, Soudleyan Stage), Whittery Quarry, Whittery Wood, Chirbury, Shropshire (SO 2746 9808). Figured Whittard 1961, pl. 24, fig. 15. GSM 87169.

2. Cephalon with parts of five attached thoracic segments, internal mould. Dorsal view. × 3. Note weak 1p furrows. Meifod, Montgomeryshire. NMW 62.24G.42.1.

3. Free cheek, internal mould, dorsal view. × 3. Chatwall Flags (Caradoc, Soudleyan Stage), Willstone, Shropshire (SO 4858 9512). GSM Mi401.

4. Free cheek, internal mould, dorsal view. × 3. Horizon and locality as fig. 3. GSM Mi403.

5. Pygidium, internal mould, dorsal view. × 4. Chatwall Flags (Caradoc, Soudleyan Stage), Onny River, 2130 yd at 284° from Wistantow church (SO 4132 8610). GSM FGD1144.

6. Cranidium, partly exfoliated, dorsal view. × 6·5. Note striated sculpture. Dufton Shales, *corona* Beds (Caradoc, Lower Longvillian), Harthwaite Sike, Dufton, Westmorland (NY 7070 2473). GSM PJ3617.

7. Cranidium, internal mould, dorsal view. × 6. Note weak 1p and 2p furrows. Gaerfawr Grits (Caradoc, Soudleyan Stage), quarry at W end of Moel-y-Garth Hill, 2–2½ miles NW of Welshpool, Montgomeryshire. ICMM 5359.

8. Incomplete cranidium, internal mould, dorsal view. × 5. Note three pairs of weak lateral glabellar furrows. Holotype of *Proetidella? marri* Dean, 1962. Horizon and locality as fig. 6. Figured Dean 1962, pl. 16, fig. 4 and pl. 17, fig. 8. BM In54644.

Decoroproetus jamesoni (Reed, 1914)

9a–c. Holotype, complete specimen; dorsal views of: 9a silicone rubber cast of incomplete external mould and 9b complete internal mould. 9c: lateral view of internal mould. All × 5. Balclatchie Group (Caradoc), Balclatchie, near Girvan, Ayrshire. Figured Reed 1914, pl. 4, fig. 8. BM In21971.

10a, b. Silicone rubber cast of external mould of cephalon; 10a dorsal view, 10b lateral view. Both × 5. Holotype of *Proetus trefoileum* Begg, 1951. Figured Begg 1951, pl. 1, fig. 2. Horizon and locality as fig. 9. HM A4123.

12. Cranidium with external surface preserved, dorsal view. × 9·5. Holotype of *Proetus vicinus* Reed, 1940. Figured Reed 1940, pl. 8, fig. 1. Horizon and locality as fig. 9. BM In37547.

13a, b. Incomplete cephalon with parts of four attached thoracic segments; 13a dorsal view, 13b lateral view. Both × 5. Holotype of *Proetus balclatchiensis* Begg, 1951. Figured Begg 1951, pl. 1, fig. 1. Horizon and locality as fig. 9. HM A4122.

11. Cranidium, internal mould, dorsal view. × 9·5. Holotype of *Proetus ardmillanensis* Begg, 1947. Figured Begg 1947, pl. 3, fig. 3. Balclatchie Group (Caradoc), Dow Hill, near Girvan, Ayrshire. HM A3692.

Decoroproetus jamesoni? (Reed, 1914)

47

14. Free cheek, internal mould, dorsal view. × 20. Upper Stinchar Limestone (Caradoc), old quarry 600 yd WNW of Kirkdominae Ruins, Stinchar Valley, near Girvan, Ayrshire (NX 249 939). Figured Tripp 1967, pl. 2, fig. 14. HM A6781a.

15. Cranidium, internal mould, dorsal view. × 17·5. Horizon and locality as fig. 14. Figured Tripp 1967, pl. 2, fig. 13. HM A6780.

16. Cranidium, internal mould, dorsal view. × 16·5. Upper Stinchar Limestone (Caradoc), Aldons Limeworks, Stinchar Valley, near Girvan, Ayrshire (NX 198 896). Figured Tripp 1967, pl. 2, fig. 15. HM A6782.

(Continued on previous page)

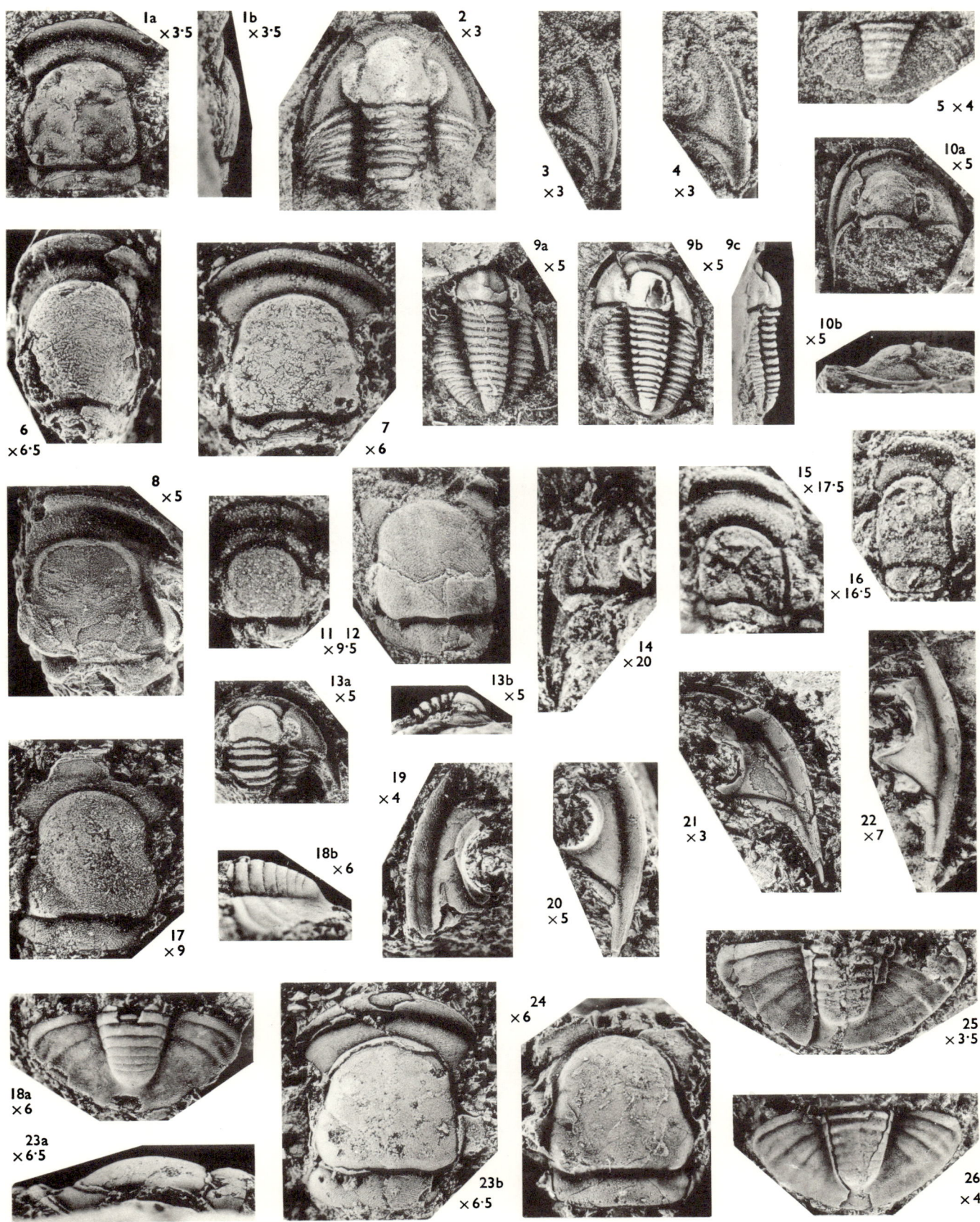

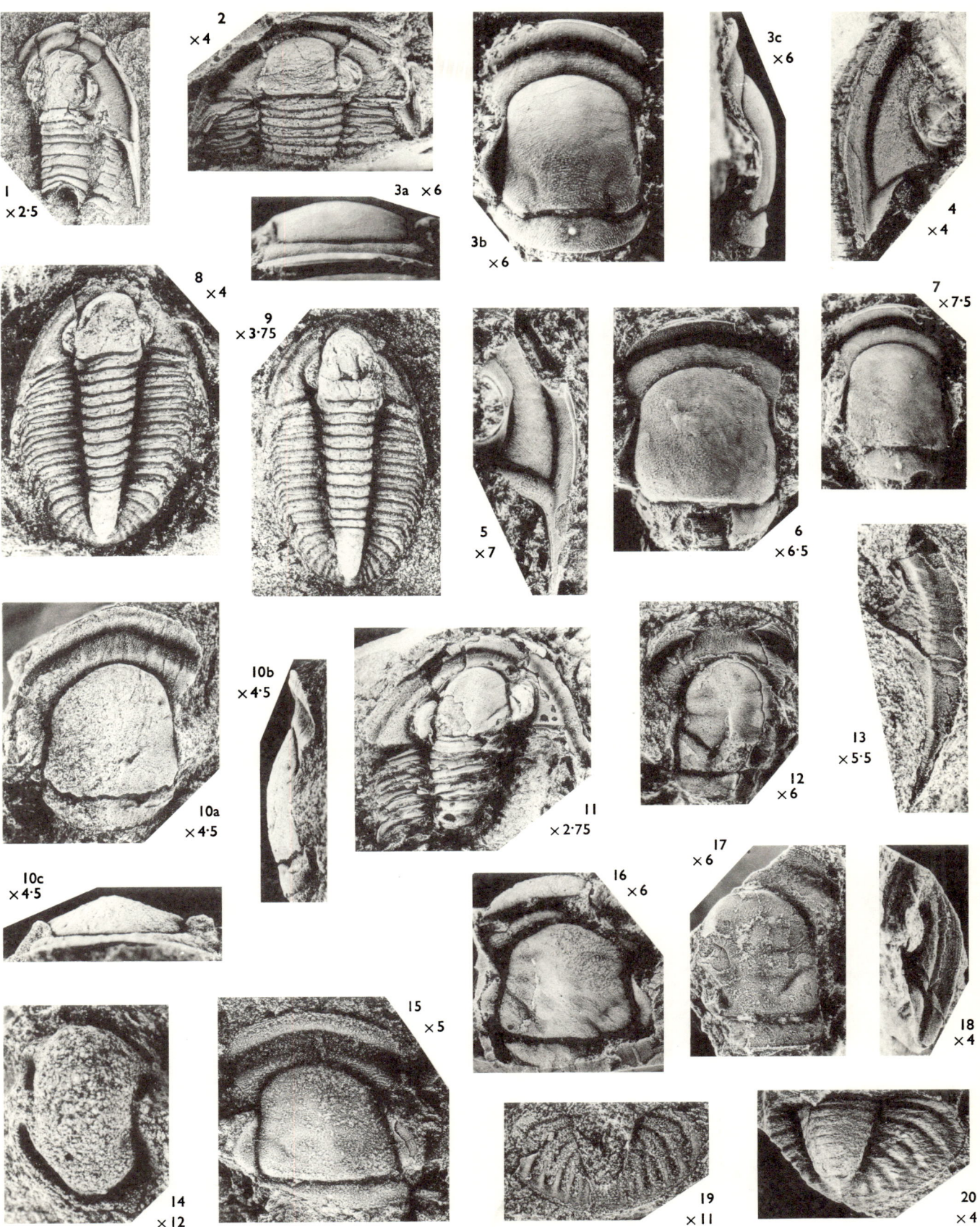

PLATE 10

OWENS, Proetid Trilobites

Plate 10

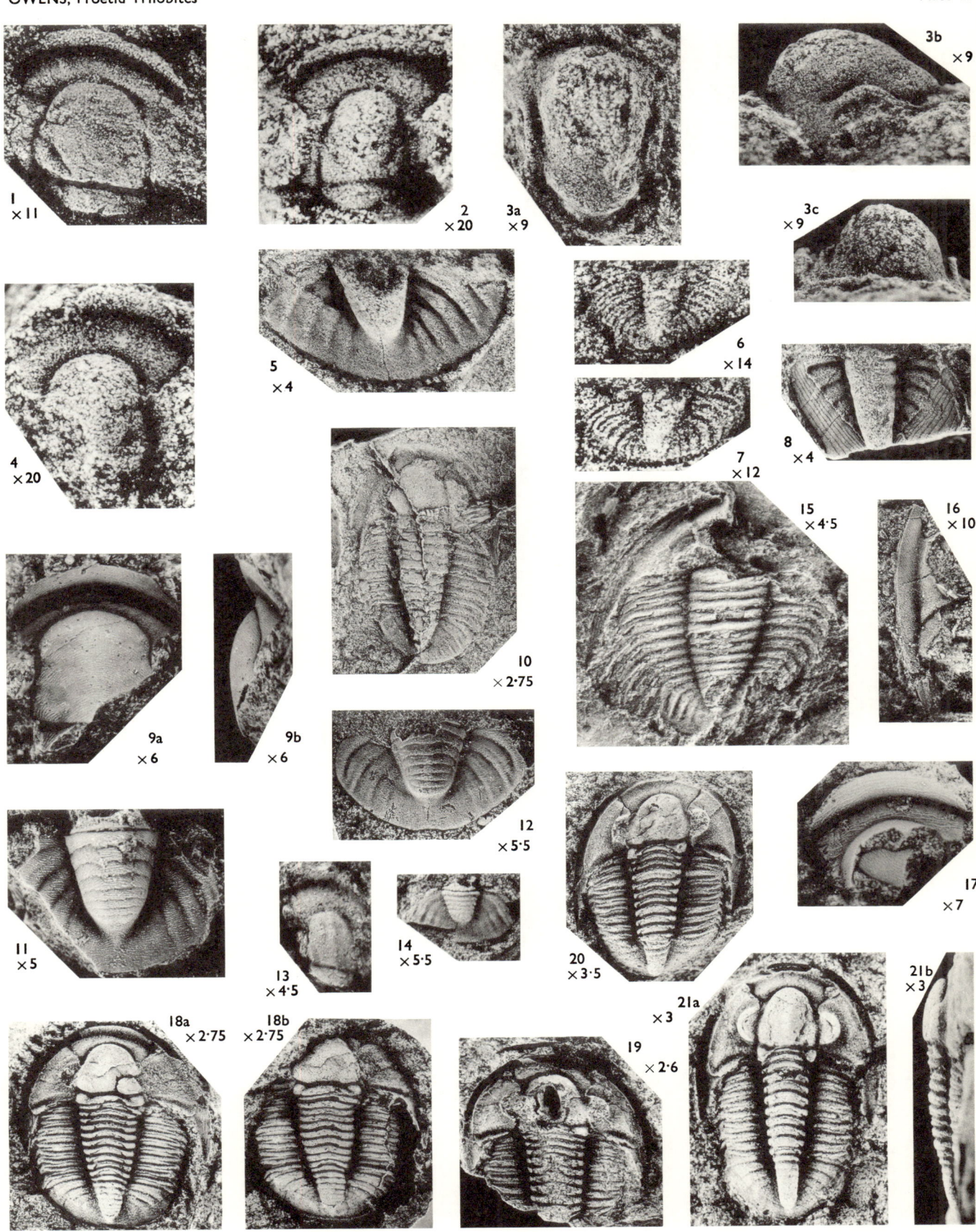

PLATE 11

Plate 11

OWENS, Proetid Trilobites

Plate 12

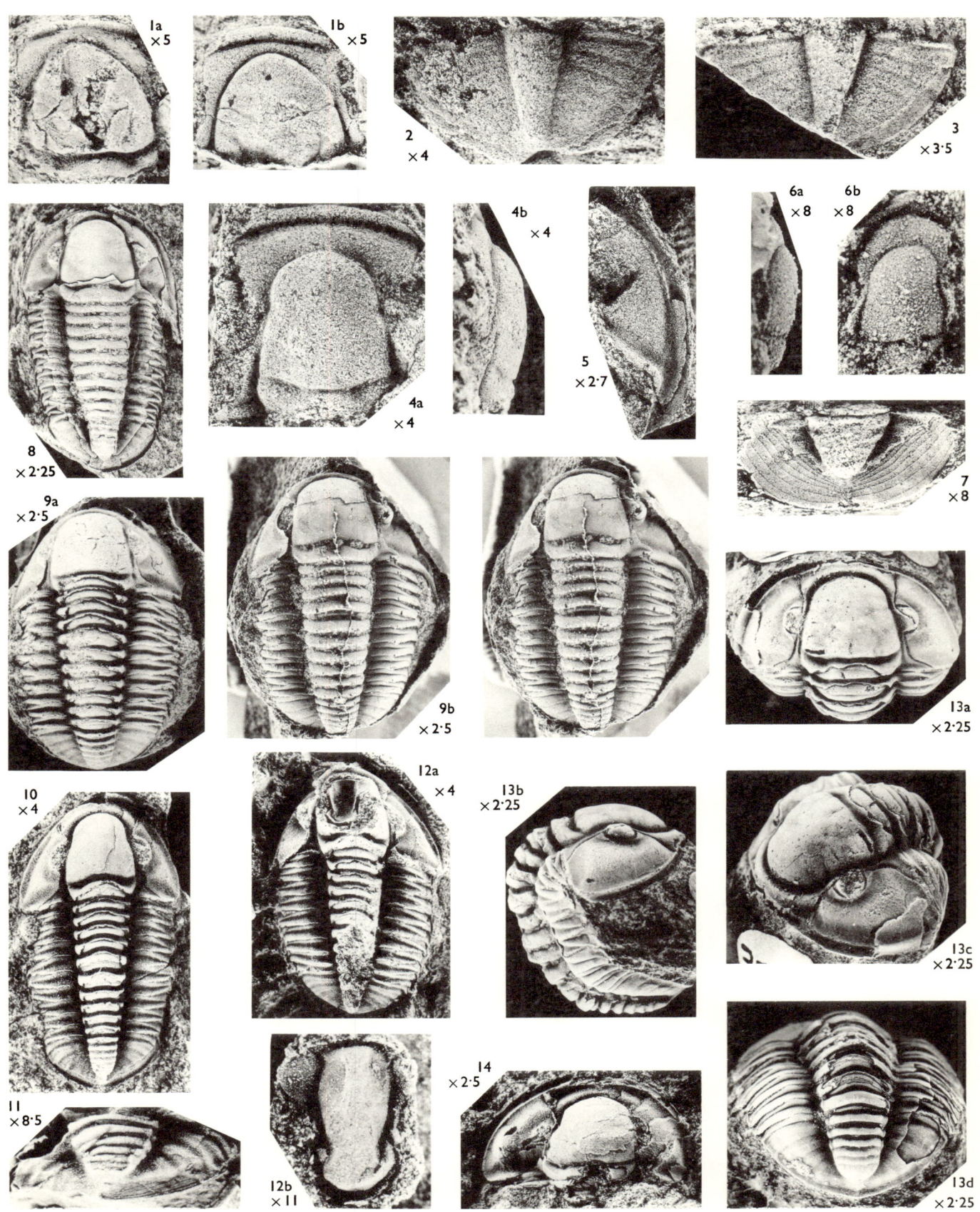

PLATE 13

Fig.

Paraproetus girvanensis (Nicholson & Etheridge, 1879)

1. Internal mould of cephalon with parts of seven thoracic segments and external mould of cephalon with parts of three thoracic segments, dorsal view. ×3·5. Ashgill mudstones (probably Rawtheyan), Dynanau farmyard, Llanystumdwy (SH 481 396). BM It8857.

2. Cephalon, internal mould, dorsal view. ×4. Horizon and locality as fig. 1. BM It8856.

3. Silicone rubber cast of internal mould of cranidium, dorsal view. ×9. *Dalmanitina mucronata* Beds (Ashgill, Rawtheyan Stage), SW of Torver Beck, Lancashire (SD 2756 9613). SM A43155.

4a, b. Complete internal mould; 4a pygidium and thorax, dorsal view, 4b cephalon and part of thorax, dorsal view. Both ×2·5. Upper Drummuck Group (Ashgill, Rawtheyan Stage), Thraive Glen, near Girvan, Ayrshire. Figured Reed 1904, pl. 11, figs. 2, 2a. BM In21914.

Warburgella (*Warburgella*) *stokesii* (Murchison, 1839)

67

5a–c. Proposed neotype, complete specimen; 5a dorsal stereograph, 5b lateral view, 5c anterior oblique view. All ×4. Wenlock Limestone, Dudley, Worcestershire. Figured Whittard 1938, pl. 3, fig. 1. BU 176.

6. Complete specimen, dorsal view. ×4·75. Horizon and locality as fig. 5. SM A28262.

7. Complete specimen, dorsal view. ×3. Horizon and locality as fig. 5. SM A28256.

8. Free cheek, dorsal view. ×6·5. Note double tropidium on posterior part of field. Wenlock Limestone, Clenchers Mill, Ledbury, Herefordshire (SO 735 376). LRU 53760.

9. Pygidium, dorsal view. ×6. Wenlock Limestone, Nodular Beds, Wren's Nest Hill, Dudley, Worcestershire. LRU 53758.

10a, b. Partially exfoliated hypostome; 10a dorsal view, 10b lateral view. Both ×9·5. Note prominent maculae. Wenlock Limestone, stream section NE of Greenpool farm, N of Coed-y-paen, Usk, Monmouthshire (ST 3343 9991) [locality now flooded by Llandegfedd Reservoir]. Figured Reed 1916, pl. 8, fig. 9, SM A16600.

11. Incomplete cranidium, dorsal view. ×7. Note preglabellar ridge. Horizon and locality as fig. 8. LRU 53759.

12. Free cheek, dorsal view. ×4. Horizon and locality as fig. 10. Figured Reed 1916, pl. 8, fig. 8. SM A16598.

13a, b. Cranidium; 13a lateral view, 13b dorsal view. Both ×9. Note preglabellar ridge. This specimen and those illustrated in figs. 10 and 12 are syntypes of '*Proetus stokesi* var. *bellula*' Reed, 1916. Figured Reed 1916, pl. 8, fig. 6. SM A16596.

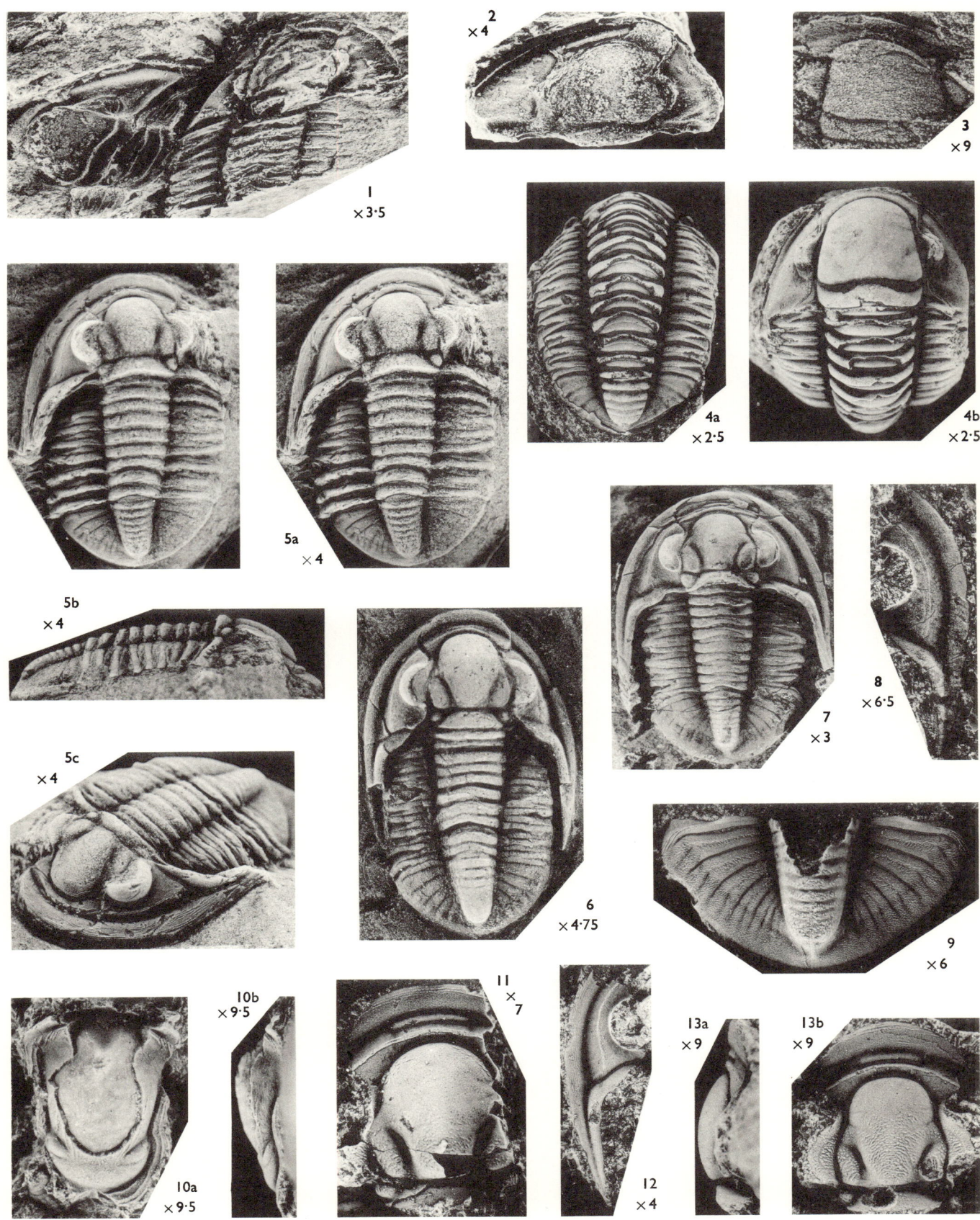

PLATE 14

Plate 14

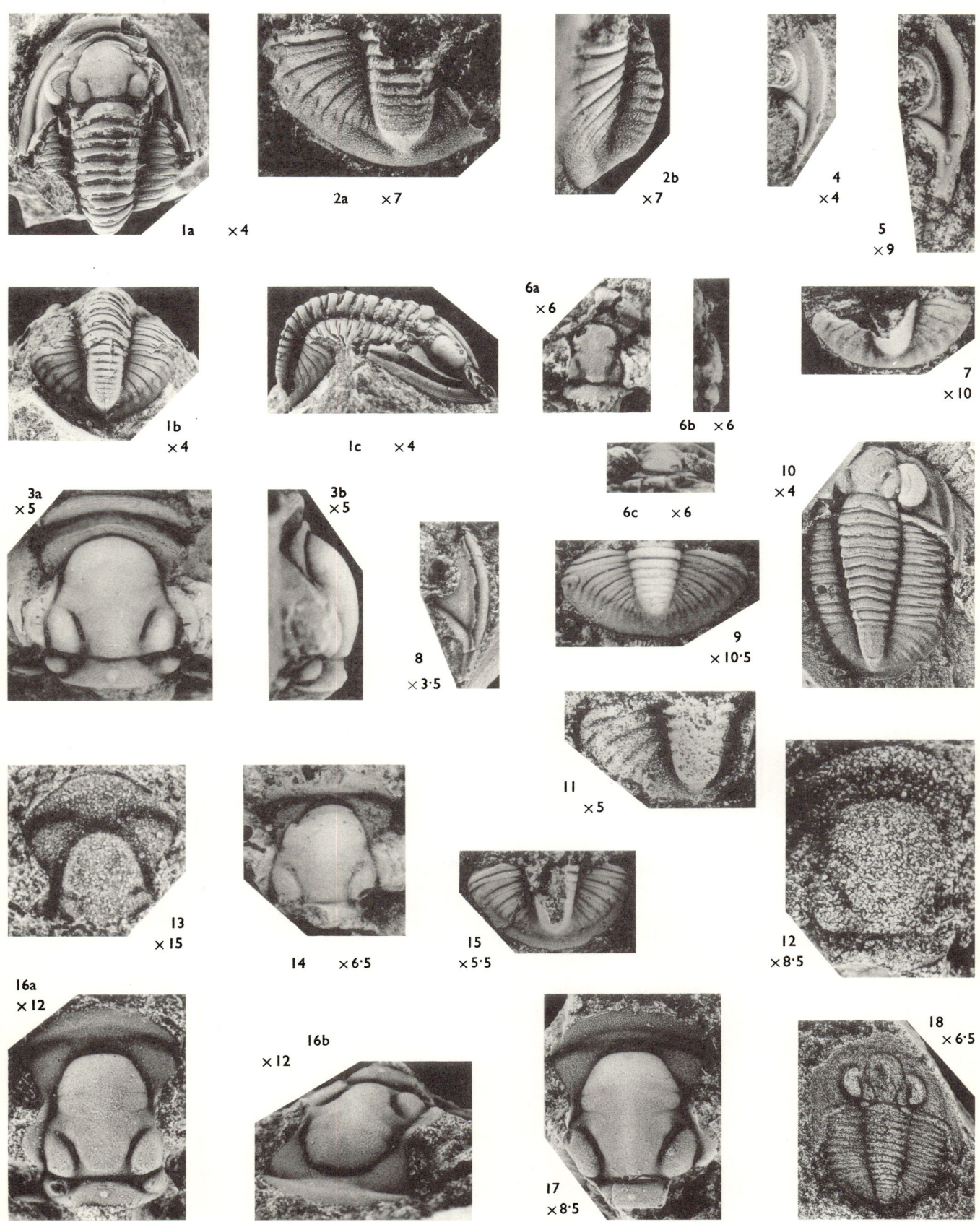

PLATE 15

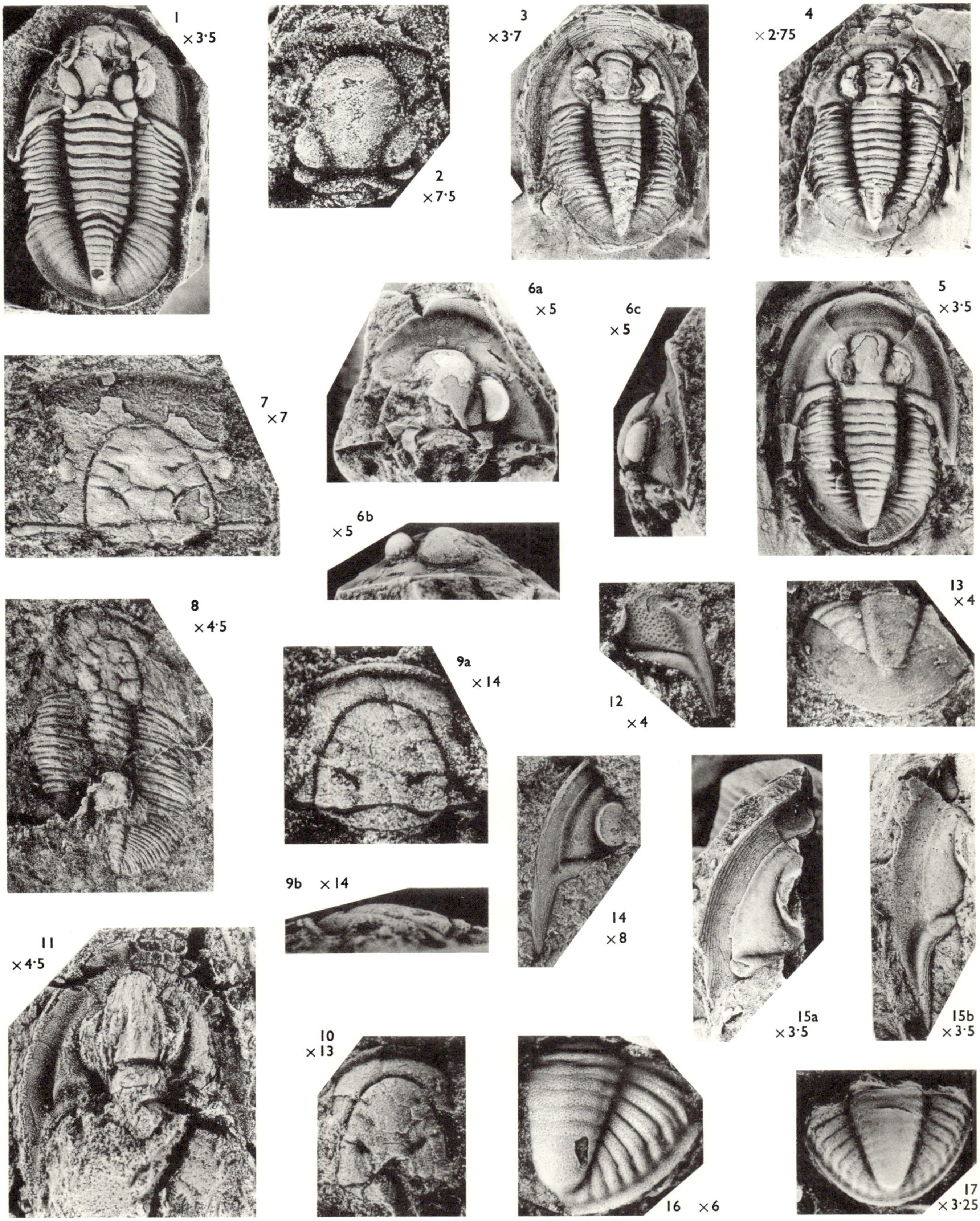

1 ×3·5

2 ×7·5

3 ×3·7

4 ×2·75

6a ×5

6c ×5

5 ×3·5

7 ×7

6b ×5

8 ×4·5

9a ×14

12 ×4

13 ×4

11 ×4·5

9b ×14

14 ×8

15a ×3·5

15b ×3·5

10 ×13

16 ×6

17 ×3·25

Cotswold Collotype Co. Ltd., Glos.